Virtual Worlds as Philosophical Tools

Virtual Worlds as Philosophical Tools

How to Philosophize with a Digital Hammer

Stefano Gualeni
University of Malta

palgrave
macmillan

First published 2015 by
PALGRAVE MACMILLAN

Palgrave Macmillan in the UK is an imprint of Macmillan Publishers Limited, registered in England, company number 785998, of Houndmills, Basingstoke, Hampshire RG21 6XS.

Palgrave Macmillan in the US is a division of St Martin's Press LLC, 175 Fifth Avenue, New York, NY 10010.

Palgrave Macmillan is the global academic imprint of the above companies and has companies and representatives throughout the world.

ISBN: 978–1–137–52177–4

This book is printed on paper suitable for recycling and made from fully managed and sustained forest sources. Logging, pulping and manufacturing processes are expected to conform to the environmental regulations of the country of origin.

A catalogue record for this book is available from the British Library.

Library of Congress Cataloging-in-Publication Data

Gualeni, Stefano, 1978–
Virtual worlds as philosophical tools : how to philosophize with a digital hammer / Stefano Gualeni, University of Malta.
pages cm
ISBN 978–1–137–52177–4
1. Virtual reality. 2. Philosophy. 3. Human behavior models. 4. Human-computer interaction. I. Title.

BD331.G765 2015
102.85—dc23 2015013138

This book is dedicated to that one cat I saw when I was five years of age.

Contents

List of Figures

From the mediatic turn to *Gua-Le-Ni*

Foreword by Jos De Mul, Erasmus Universiteit Rotterdam

In the last couple of decades a new discipline called media philosophy has entered the philosophical arena. According to Reinhard Margreiter, one of its proponents, the name media philosophy refers not merely, nor even predominantly, to the exploration of yet another ontological domain, but rather designates a fundamental transformation of philosophy itself. This fundamental transformation is characterized by a turn toward (the descent and history of) the mediatic foundations of philosophy. In his view, media philosophy might become a contemporary "*prima philosophia*" (Margreiter 2003, 151). Margreiter does not argue, however, for a modernist kind of foundationalist superdiscipline, but for a critical discourse that has to accompany every act of knowing.

Though the name media philosophy is a recent invention, the phenomenon is not altogether new. Fundamental reflections on the impact of writing on philosophy can already be found in Plato's *Phaedrus* and *Seventh Letter*, that is, on the type of oral philosophy that preceded written philosophy and that is still reflected in the dialogical form of Plato's writings. However, in the tradition of Western philosophy, which is strongly connected with the medium of the book, these kinds of reflections remained relatively scarce and marginal for a long time. Starting from Parmenides' identification of being and thinking, a dominant part of the metaphysical tradition was based on the presupposition that thinking and being – *nous* and *phusis* – share the same form (*eidos, morphé*), guaranteeing the identity of what can be thought and what can be (cf. Allen 2004, 218).

Kant's transcendental philosophy can be regarded as the first radical critique of the outlined metaphysical equation of thinking and being. According to Kant there can, in fact, be no immediate and absolute knowledge of reality, as that human knowledge depends on the finite *medium* of our faculties of sensibility, understanding, and reason. However, as Kant deems this medium to be timeless and shared by all human beings, he could still adhere to the notion that the phenomenological world constituted by this medium is something that has empirical objectiveness and, as such, is open to scientific explanation, prediction, and control. In post-Kantian philosophy two further developments can be distinguished, which together have resulted in what might be called the mediatic turn in modern philosophy.

The first of these developments has to do with the *historicization* of human reason (cf. De Mul 2004, 97–125). After Kant, the idea emerged that human reason is not a timeless entity but something that develops in – natural and historical – time. In Hegel's philosophy this historicization was still regarded as a process in which Absolute – that is, suprahistorical – Reason finally becomes conscious of itself (and in this sense returned to a pre-kantian metaphysics). However, in the hermeneutical tradition – starting with Dilthey and radicalized in Heidegger and his post-modern heirs – the emphasis gradually shifted to the finiteness of human experience and the radical historical character of ontology.

The second development in the post-kantian philosophy I refer to is what might be called the *externalization* of human reason, which emerged from the realization that the thinking of being always requires an external medium. Already in the nineteenth century, Herder and Von Humboldt emphasized both the crucial role natural language plays in thinking, and the non-transparency of this medium. In the continental philosophical tradition it was in hermeneutics – in which we might include Nietzsche, the philosopher with the hammer who blamed grammar for our belief in God (Nietzsche 1980, Band 6, 78) – that this insight was developed further.

In the analytical tradition, a similar development took place in the, so-called, "linguistic turn" (Rorty 1967). This turn was accompanied with the "belief that the problems of philosophy may be solved or dissolved either by reforming language (the advocates of this were dubbed 'ideal language philosophers') or by a better understanding of the language we actually use ('ordinary language philosophers')" (Hacker 2007). Wittgenstein played a crucial role in both manifestations of the linguistic turn. In the mainstream interpretation of Wittgenstein – sketched broadly enough to abstract it from the many disagreements – in the *Tractatus*, Wittgenstein held that "the sentences of our language, fully analyzed, necessarily reflect the metaphysical form of the world" and that "all philosophy is a critique of language." However, in his belief in the correspondence between being and the logical form of language, Wittgenstein – in spite of his radical restriction of meaningful language to elementary and complex propositions of science and his critique of every philosophy that pretends to go beyond these propositions – remained a victim of Platonic metaphysics. In his later writings – I am still sketching the mainstream interpretation – Wittgenstein criticized his earlier position and developed a therapeutic philosophy that aimed to dissolve philosophical problems by analyzing the many confusions that characterize our ordinary language.

What distinguishes recent media philosophy from the earlier continental and analytical approaches is the fact that its scope goes far beyond the linguistic domain. Inspired by the emergence and impact of new media (radio, film, television, and the computer) and by the work of otherwise diverse thinkers, such as Cassirer, Langer, McLuhan, Ong, Goodman and Derrida, to mention a few, the mediatic self-reflection has been extended to (the symbolic and material dimension of) all cultural media of experience. In this mediatic turn the development of computer mediation in the second half of the twentieth century has become a central topic. One of the reasons for our fascination with computers is that, with the development of information and communication technologies (ranging from artificial intelligence and expert systems, to social media and Big Data analysis), the outsourcing and supplementation of human reason seem to enter an entirely new phase, which in its radicalism perhaps can only be compared to the externalization of thinking in writing, several millennia ago.

Stefano Gualeni's *Virtual Worlds as Philosophical Tools: How to Philosophize with a Digital Hammer* offers a highly original and recommendable contribution to the mediatic turn. In this interdisciplinary investigation he combines the theoretical insights of philosophy and media studies with his practical experience as a game designer. It is the ludic perspective of his work which makes this study especially worth reading. Of course, phenomena such as computer games, serious learning, and the "ludification" of human identity and culture has attracted much scholarly attention in the last couple of decades (cf. Frissen et al 2015). What distinguishes Gualeni's study is his penetrating analysis of the ontological dimension of computer games. Viewing man and world *sub specie ludi* is of course not an entirely new phenomenon. Already early on in Western thought, Heraclitus speculated that "the course of the world is a playing child moving figures on a board – the child as absolute ruler of the universe" (Sprague 2001, 26). And more recently, in his highly influential *Homo ludens* (first published in Dutch in 1938), Johan Huizinga claims that play is the very origin of all human culture: "It does not come from play like a babe detaching itself from the womb: *it arises in and as play, and never leaves it*" (Huizinga 1955,

173, italics JdM). Huizinga argues that this is not only true for cultural phenomena like rituals, sports, theatre, legal practice, and warfare, but is no less so for the philosophical attempts to understand the world with the help of concepts. Echoing Kant and Schiller, Huizinga emphasizes the crucial role of playful human imagination.

Gualeni's original contribution in his study is that he shows that the development of computer games has created an exciting new domain for philosophical imagination, by disclosing a digital space for exploration and experimentation. I am certain that the readers who follow Gualeni on his ludic path will not only experience that this study is a fun read, but that it will also expand their ontological horizon. Well, if that isn't serious gaming, what else is?

References

Allen, B. 2004. Wittgenstein's Onto-Logics. In *Post-analytic Tractatus*, (edited by) B. Stocker. Aldershot, Hants, England; Burlington, VT: Ashgate.

De Mul, Jos. 2004. *The Tragedy of Finitude. Dilthey's Hermeneutics of Life*. New Haven: Yale University Press.

Frissen, V., De Lange, M., De Mul, Jos., Lammes S. and Raessens, J. (eds). 2015. *Playful Identities. The Ludification of Digital Media Cultures*. Amsterdam: Amsterdam University Press.

Hacker, P. 2007. Analytic Philosophy: Beyond the linguistic turn and back again. In *The Analytic Turn: Analysis in Early Analytic Philosophy and Phenomenology*, (edited by) M. Beaney. London: Routledge.

Huizinga, L. 1955. *Homo Ludens: A Study of the Play-Element in Culture*. Boston, MA: Beacon Press.

Margreiter, R. 2003. Medien/Philosophie ein Kippbild. In *Medienphilosophie: Beiträge zur Klärung eines Begriffs*, (edited by) S. Münker, A. Roesler and M. Sandbothe. Frankfurt am Main: Fischer Taschenbuch Verlag.

Nietzsche, F. 1980. *Sämtliche Werke. Kritische Studienausgabe*. 15 vols. Berlin: De Gruyter.

Sprague, R. K. (ed.). 2001. *The Older Sophists: A Complete Translation by Several Hands of the Fragments in Die Fragmente Der Vorsokratiker*, (edited by) Diels-Kranz. With a new edition of Antiphon and of Euthydemus. Indianapolis, IN: Hackett Publishers.

Preface and Acknowledgments

What is it like to be a human being in a simulated world? Will experiencing worlds that are not "actual" change our way of structuring thought? Can virtual worlds open up new possibilities for philosophizing? *Virtual Worlds as Philosophical Tools* tries to answer those questions from a perspective that is informed and inspired by the philosophy of technology, media theory and the design of videogames. Despite being presented here in a form that is almost exclusively textual, its contents encapsulate the interdisciplinary work of a digital humanist and, as such, are characterized by a degree of practical involvement in the creation and repurposing of digital technology and digitally mediated contents. More specifically, this book emerges from the engagement "in design and development processes that give rise to richer, multidirectional models, genres, iterations of scholarly communication and practice" (*Digital Humanities Manifesto 2.0*, available online at http://www.humanitiesblast.com/manifesto/Manifesto_V2.pdf, page 6).

For the most part, the ideas discussed in *Virtual Worlds as Philosophical Tools* matured in relation to my passion for videogames and to my professional activity as a videogame designer. It was my involvement in the creation of virtual worlds that originally motivated many of the theoretical notions and philosophical perspectives presented in the book, as well as the idea of turning philosophical notions and thought experiments into concrete (and often playful) simulations. Materialized in the form of virtual worlds, philosophical perspectives and concepts can be objectively encountered and manipulated by both the recipients and the crafters (the designers) of the philosophical experience, *de facto* opening up new, experiential and experimental possibilities for philosophy. In that sense, virtual worlds are understood as the context in which new ways of pursuing the humanities have already begun to arise.

The secondary title of this book is a playful reference to Nietzsche's 1888 *Twilight of the Idols, or, How to Philosophize with a Hammer*. In Nietzsche's book, his metaphorical hammer materializes the force (that of his own ideas) with which he intended to "sound out" the "idols" of religion and traditional beliefs. By smashing these with his philosophical hammer, Nietzsche's philosophical goal was to reveal them as hollow and false. In my work, I embrace the metaphor of the hammer in a way that is significantly different and more instrumental than Nietzsche's; in this book, the digital hammer represents the ways in which our perceptions and our cognitive and operational capabilities are extended by the experience of interactive virtual worlds.

The choice for the title might suggest that this book develops philosophical arguments by adopting a traditional, instrumentalistic perspective. I would like to clarify, right from the start, that I do not consider virtual worlds to be neutral tools, nor am I advancing the claim that technologies are tools over which we have complete mastery (both in their uses and in their social employment). From my perspective, technologies are never impartial instruments to be employed for good or bad purposes. Accordingly, in this book, virtual worlds are not understood as simply and discretely supplementing our perceptual, cognitive, and operational limitations; all technological artifacts that assist us, constrain us, co-shape who we are, are inevitably integral components of our social, ethical, and intellectual horizons. In other words, we are not shaping our thinking and who we are through a mere instrumental use of technologies, but we are shaping our technical horizon, ourselves, and our thought as part of the same cultural process. Put simply, this book understands all technological systems and artifacts as mediators. In particular, it presents the virtual worlds disclosed by videogames and simulations as tools that afford us the possibility for shaping our thoughts and our behaviors in contexts that are less inflexible and experientially homogenous than the world that we "actually" share as biological organisms.

In that respect, the American psychologist Abraham H. Maslow famously observed that even the brandishing of a simple tool has the potential to influence not only our operational capabilities, but also our cognitive structures. To a man holding a hammer, to paraphrase Maslow, everything looks like a nail. Analogous to the ways in which an actual hammer both extends and constrains our possibilities for thinking about the world and acting within it, in this book I understand virtual worlds as very peculiar ways in which human thought and human agency can be mediated.

Next to the deliberate, functional use of technologies and technological artifacts to complement our existence, the very design of technologies and technically mediated content can also be understood as a transformative "technology of the self" (a practice that assists individuals in objectifying and transforming themselves and their relationship to the world). As will be discussed specifically in the Chapter 4 of this book, any technological design also contributes to shaping the subjectivity of its designers.

Embracing the digital-humanistic perspectives outlined above, I wrote this book with two objectives in mind:

1. to provide an ontological account of what "everything looks like" to our digital-hammer-wielding culture (a hermeneutical perspective);
2. to explore the expressive limitations and possibilities inherent in using virtual worlds specifically as philosophical mediators (that is to say, as ways to materialize philosophical concepts, perspectives, and thought experiments) and as "technologies of the self".

Virtual Worlds as Philosophical Tools has been in development for about six years. During its conceptualization and refinement, many of the notions and perspectives that are articulated in this book were discussed in a variety of formats, including academic journals, videogame developers' conferences, public lectures, academic conferences, and a doctoral dissertation. Having already acknowledged my debt to videogames as both designer and player (for having shaped videogames and for having been shaped by videogames), I would like to dedicate a few lines to expressing my gratitude to the individuals, communities, and institutions that supported and challenged me through the process of conceptualization and writing.

First and foremost, I wish to acknowledge Professor Jos De Mul, whose careful guidance made this book possible. I am sure that I will never be able to express sufficient gratitude for Professor De Mul's humanity and support.

I am grateful to NHTV Breda University of Applied Sciences (the Netherlands) for having provided many of the logistical and economic resources that were necessary during the pursuit of this phase of my philosophical project. I would like to thank the Laguna College of Art and Design (California, United States), and in particular Sandy Appleoff, for the enthusiasm with which they have always embraced and supported my work. I would also like to express my appreciation to the University of Malta for believing in me and my project, and for giving me the opportunity to continue to work on both. Finally, a few words of appreciation go to the academic community of the Philosophy of Computer Games for graciously and constructively challenging and fostering the interdisciplinary and practice-infused perspectives that are presented in the pages that follow.

1
The Questions Concerning Digital Technology

1.1 What is technology?

In ancient Greece, the word *τέχνη* (*techné*) was used largely to indicate the attitude, the methodology, or the skill aimed at the practical creation of a material thing. This practice-oriented concept was later appropriated by the Romans under the umbrella term *ars* (art) (Fedier, 2001, 12–27). In English, the attitude towards "making" that the Greeks identified with the word *techné* is commonly translated as "craftsmanship," "technology," or "art."

Martin Heidegger was a German thinker whose pioneering philosophical insights into technology had a foundational role in the structuring and development of the perspectives and ideas discussed in this book. In his 1938 essay "The Age of the World Picture," Heidegger criticized the modern understanding of *techné*, finding the interpretation unfaithful to the original meaning. Heidegger explained that, in ancient Greece, the word *techné* never signified the action of making, but rather indicated an epistemological approach – a perspective capable of revealing the world in a specific "light." Correspondingly, Heidegger (2008) described works of art as artifacts endowed with the potential to disclose worlds, that is to say, to open up new ways in which reality can "unconceal" itself. Although he never explicated in detail how he thought art could be capable of engendering such disclosure, Heidegger identified experiential influences in the way people structured their relationships with reality (and thus allowed for the emergence of "worlds") as the sole cultural role of artistic production. In the final phase of Heidegger's thought, he extended to all things the capability to let "things come into being" and to open up new worlds and worldviews, previously attributed only to artworks (Verbeek, 2005, 89).

Since the word *techné* was coined, the notions of craftsmanship, technology, and art have gradually developed into separate contexts. These contexts do, however, intersect in multiple ways and combinations. Dutch philosopher Jos De Mul observed that, in Western culture, the creations of craftspeople and artists have always depended on the mastery of specific productive or expressive tools, and contemporary artists are no less reliant on technological tools than were their prehistoric predecessors (De Mul, 2010, 139). The modern dependence of artistic and cultural production on technological mediation appears particularly obvious in the creation of content for digital media. The creative instruments afforded by computers disclose a vast horizon of combinatorial possibilities for expression and interaction – possibilities that are completely dependent on their technological platforms.

The historical shift and fragmentation of the meaning of *techné*, together with the baffling diversity of the possible interpretations of the word "technology," inspired Stephen J Kline's 2003 essay, "What is Technology?" Kline clarified that the term "technology," rather than representing a single concept, currently refers to a variety of concepts bound together by the common characteristic of creating or employing man-made objects. According to Kline, the depth and intricacy of humankind's involvement with technology is the reason for the ambiguous nature of the term as it is currently used. To demonstrate this ambiguity, the word "technology" can mean:

- Anything that the natural environment does not generate without human intervention, such as "refrigerators, eyeglasses, atom bombs, paints, automobiles, pianos, paper, rubber, glass, aspirins, penicillin, airplanes, copying machines, furniture, roads, rifles, printing presses, boots, bicycles and on and on" (Kline, 2003, 210).[1]
- All the expertise and methodologies employed in the pursuit of a practical task of some kind. This description of "technology" can be identified as the closest to the original meaning of the Greek word.
- All the constituent parts required to produce different types of hardware, "including its inputs: people; machinery; resources; processes; and legal, economic, political, and physical environments" (Kline, 2003, 210, 211).
- The complex social and technical infrastructures through and for which artifacts are made and used, in addition to the techniques used by individuals to produce or employ these artifacts. In this interpretation, a car, for instance, cannot be considered isolated from the machinery that is used in the car's construction or from the technical

limitations of this machinery. The way cars are built also relates to larger, interconnected technological systems such as roads and gas stations, and to traffic rules, laws regarding ownership, and the tastes and needs of the social group(s) for whom they are produced (Van den Berg, 2009, 23). The necessary relationships and dependencies among all these systems constitute the "socio-technical system of use." When using a car, "[we] use the combined system (the autos plus all the rest) to extend the human capacity for moving ourselves and our possessions about" (Kline, 2003, 211).

- A combination of natural forces conveyed and combined toward certain human purposes. In this conception of technology, the successive stages of technological development can be interpreted as one of the objective externalizations of the historical process of self-understanding (Coolen, 1992, 250–271; De Mul, 2010, 113). In line with this idea, Heidegger's 1954 essay "The Question Concerning Technology," discussed the potential of different techniques to unveil different ways of being-in-the-world.

This book explores the role of *techné* (focusing specifically on its digital manifestation) as an influential factor in socio-cultural change. More specifically, *Virtual Worlds as Philosophical Tools* articulates an understanding of virtual worlds as capable both of mediating philosophical thought and of experientially fragmenting and augmenting the ways in which people can think, perceive, and operate, expanding the boundaries beyond the mere "actual" and extending into what is virtually "possible." In structuring these two intricately related perspectives, I was motivated to adopt an interpretation of "technology" that could embrace interactive, digital worlds as constitutive components of the experience of being human. Also, I found it necessary to adopt an understanding of what "technology" means that would be suitable for encompassing most of the interpretations described above. With these objectives in mind, I drew on De Mul's work, where technology is defined as "a conglomerate of…artefacts, specific forms of knowledge and capabilities…(embraced in their necessary relation with the relative) geographical and social infrastructure, economic interests and societal norms and values" (De Mul, 2002, 30).

1.2 A philosophical task?

In *The Metaphysics of Virtual Reality,* Michael Heim maintained that the way computers produce interactive virtual environments and allow smooth and controlled transitions "to the real and back" cannot be satisfactorily framed with models and analogies used to analyze traditional forms of mediation or psychotropic experience. Rather, the profound cognitive, epistemological, and sociological implications of interaction with virtual worlds necessitate philosophical exploration and understanding. In line with Heim, I believe that, in the age of digital mediation, it is hard to imagine a *more* philosophical task than reflecting on how virtual experiences affect what it is like to be a human being. Another, and more Heideggerian, way to express the core motivation of this book is to phrase the main matter of exploration as a simple question: How can interactive digital technology assist people in "overcoming" the traditional boundaries of human ontologies?

This guiding question may be read by some as a provocative paradox; from a traditional humanistic perspective, the mechanization of the world and the progressive penetration of technology into social processes and practices are not customarily understood as conducive to people overcoming their limitations or being in any way liberated. Collectively, these processes are, in fact, more often dreaded as a force that constantly challenges and threatens the values that make up the fabric of human society. This perspective (known as the technologically deterministic standpoint), often paired with the perception of impending threats inherent in people's increasing dependence on technology, was popularized by the paradigmatically techno-pessimistic cultural production of the early twentieth century.

Mary Shelley's 1818 novel, *Frankenstein; or, the modern Prometheus,* is often recognized as a precursor to the use of fictional media to reinforce and spread the ideology that the interference of technology on a favorable balance of natural and social forces will inexorably lead to dehumanizing and tragic consequences for mankind. One example of this later techno-pessimistic cultural production is Edward Morgan Forster's 1909 novel, *The Machine Stops*, which highlights the escalating detachment of humankind from the world through a complete and alienating dependence upon technology. In the artificial underground environment of *The Machine Stops*, machine technology is ubiquitous, inscrutable, indispensable, and revered in a quasi-religious sense. Only upon the final failure of the machine do the characters realize how far removed they have become from the (naively idealized) natural and social order of which they were once a part (Forster, 1985). Another remarkable early twentieth century example of techno-pessimism in cultural production is *R.U.R.* (*Rossum's Universal Robots*), a theatrical play created in 1920 by Czech playwright Karel Čapek. *R.U.R.* is often considered a milestone in science-fiction, because the play introduced the term *robot* (Czech for work) to indicate an electro-mechanical agent capable of pursuing tasks autonomously or semi-autonomously, and because it established the popular culture trope of the deliberate revolt of the machines against their creators.

From the technologically deterministic, techno-pessimistic perspectives outlined above, being human is understood as an involvement with reality that is genuine and irreducible. Technology is, instead, dialectically recognized as an autonomous force striving to dominate rationally a world made of objects, including mankind (Heidegger, 1982; Vattimo, 1991, 40, 41; Richard Villa, 1996, 182; Costa, 2007, 33–47). In this traditional, humanistic approach to the philosophy of technology, no form of technical mediation could be interpreted as contributing to society and culture through the emancipation of humanity from cognitive and operational limitations or, in fact, to the emancipation of humanity in any form. Rather, in this perspective, technical mediation would, instead, manifest itself as the materialization of the will to control and rationally reconstruct the world recorded, for example, in social science-fiction literature or in some of Borges' fictional writings (Borges, 1994, 2001, 2004; Richard Villa, 1995, 182).

Is technology, then, to be understood as a danger for humanity? If so, what is de-humanizing about technology? Could it not be embraced, instead, as an apical form of humanism? These questions were raised by Heidegger, explicitly in the context of philosophy of technology, starting from 1949. In his writings of that period, Heidegger acknowledged the objectification of both the world and human beings as the supreme danger of advancing mechanization. Along the same line of though, he openly identified technology as the ultimate incarnation of Western thought. (Heidegger, 1982; Vattimo, 1991, 177–179; Richard Villa, 1996, 181–195).

Philosophers and aesthetics scholars of the last century (including Heidegger's student, Hans Robert Jauss) reacted to the progressive commoditization and alienation of human existence that they observed in the technical mediation of culture. They proposed, instead, the free encounter with art as a means of achieving liberation from the canons and shortcomings of our system of thought by detaching people from their everyday, functional existence and leading them into a freer realm of sensory appreciation. It was before the proliferation of computers that Heidegger, Jauss, Herbert Marcuse, and others developed their ideas concerning the social relevance of art and its salvific potential. Consequently, none of these scholars could fully anticipate the advent and the effects of a *techné* capable not only of (re)presenting fictional worlds, but also of offering persistent interactions with them.

With the objective of articulating a philosophy of technology that can frame the effects of virtual experiences on human cognition, I draw especially on Heidegger's pioneering efforts in the field of the philosophy of technology. Following Heim, I recognize in Heidegger's philosophical understanding of technology a milestone that needs to be considered, used for guidance, and critically re-thematized in the age of digital mediation. Embracing Heidegger's work as a supporting philosophical framework necessitates the clarification of some of his (notoriously obscure) lexical items that will be used frequently in structuring the arguments in this book:

(1) World

In the philosophical tradition of phenomenology, the term 'world' generally indicates a set composed of beings that are understood together with all their (detectable) properties and mutual relationships. More specifically, a world describes the set outlined above as experienced by one of the beings involved in it. To be identified as a world, such experience need to be persistently perceivable and behaviorally consistent for the being experiencing it. Those qualities make that experience emerge as an (intelligible) world for a being within a certain spatial-temporal context. Apart from being very abstract and encompassing, this functional interpretation also establishes a clear distinction between the experiences of virtual worlds and those of dreams or hallucinations. The virtual worlds of simulations and videogames are recognized as worlds precisely because they can be accessed and returned to at will, and because they emerge in ways that are repeatable and relatively stable in their mechanical and aesthetic aspects. Inspired by Heidegger's existential phenomenology, this definition of world lays the groundwork for the sidestepping of a dualistic perspective of the philosophy of mind. Instead of constructing a system of knowledge based on the theoretical separation between an observer (subject) and the world (object), Heidegger presents their coexisting and being mutually constitutive as necessary and structural aspects of a world. To put it simply, a world indicates the ways in which reality is disclosed to a being (Verbeek, 2005, 108).

(2) Ontology

The way in which a being structures its general understanding of a world is commonly referred to "an ontology". In analogy to this basic definition, the specific context of Western philosophy defines ontology as the fundamental study of the things that can be said to exist in a world, their qualities, and interrelationships. As discussed in the point above, according to Heidegger, a being (*Dasein*, German: from *da,* there, and *sein*, being) is always involved with a "there", with a world. In a general sense, I will use the unspecific term ontology to refer to human kinds of ontologies – the rational organization of a specific group of relationships constituted between a (human) being and a world. From this perspective, things in the world make sense within an ontology precisely because, via the mediation of the senses, they become part of a persistent and intelligible system of relationships with an individual being.

(3) Humanism

In his 1947 "Letter on Humanism" (re-edited for publishing in 1949), Heidegger presented an understanding of humanism that did not align with the way in which the term has commonly been used since the days of Ancient Rome. In its conventional meaning, in fact, "humanism" indicates the pursuit and the upholding of what are recognized as traditional human values (culture, art, sciences, human dignity and God) through "scholarship and training in good conduct" (Heidegger, 1998, 244–251). As a reaction to this interpretation, Heidegger explained that the conventional understanding of humanism does not truly cater to the original essence of human beings, but rather "is determined with regard to an already established interpretation of nature, history, world, and...beings as a whole" (Heidegger, 1998, 245). Heidegger found that this way of understanding humanism was a reductive by-product of the Western tradition of thought. He further specified that, in his opposition to the traditional use of the term humanism, he was not advocating for the "inhuman" or a return to the "barbaric." His opposition stemmed, instead, from the belief that humanism can only be properly understood and restored in culture as a more original way of meditating on and caring for humanity.

In the preface to this book, virtual worlds are presented as "digital hammers": as tools, or rather as mediators, that experientially allow us to transcend what is "actually present" and can contribute to the shaping of our thoughts and our behaviors in virtual contexts that are less univocal, less exclusive, and less inflexible. In structuring this perspective, I will operate within Heidegger's understanding of humanism, meaning a concern and focus on human beings that is necessarily embedded in a derivative horizon of thought and that is also capable of transcending such a horizon through remaining open to other "vistas" (Heidegger, 1998, 265).

(4) Overcoming

In this fourth linguistic specification, I would like to direct the attention of the reader to the fact that, when describing the social and cognitive effects of the exposure to interactive virtual worlds, I will not refer to these experiences as ruptures with our everyday engagement with the world. I will also not characterize the proliferation of virtual worlds and the progressive integration of digital technology with social practices as having brought about a "revolution" in our ways of thinking and operating, or as a "radical break" with humanity's pre-digital past. Instead, terms like alteration, shift, and overcoming will be employed to describe the ontological effects of the experience of virtual worlds. The last term in particular, overcoming, is used in accordance with Heidegger's embracing of the concept. Overcoming should not be understood in the dialectical meaning of the German term *Überwindung* (surpassing), but rather in the nuanced conjunction of two other terms: *Andenken* (remembrance) and *Verwindung* (distortion, twisting, incorporation). One of Heidegger's translators, Joan Stambaugh, clarified the difference between the two distinct and coexisting understandings of overcoming presented in *Being and Time*, stating that when something is overcome in the sense of being *überwunden*, it is defeated and left behind. This is not the sense Heidegger intends here. When something is overcome in the sense of being *verwunden* it is, so to speak, incorporated. For example, when one overcomes a state of pain, one does not get rid of the pain. One has ceased to be preoccupied with it and has learned to live with it. Thus, to overcome metaphysics would mean to incorporate metaphysics, perhaps with the hope, but not with the certainty, of elevating it to a new reality. (Heidegger, 1973, 84; also cfr. Heidegger, 1982, 39)

Elaborating on the same concept, contemporary Italian philosopher Gianni Vattimo maintained that overcoming as *Verwindung* "repeats [metaphysics] while radically changing its meaning" (Vattimo, 2004, 39). In his analysis, Vattimo identified two qualities of Heidegger's idea of *Verwindung*, specifically he noted that *Verwindung*

(a) consisted of a "repetition" of the metaphysical tradition, which is to be understood as an acceptance and a remembrance (*Andenken*) of it, and

(b) was a factor of change in the understanding of metaphysics itself.

Combining the two characteristic aspects of *Verwindung* in the dyadic expression "acceptance-distortion," Vattimo interpreted Heidegger's project of "overcoming" metaphysics as "a going-beyond that is both an acceptance (or 'resignation') and a 'deepening'" (Vattimo, 1991, xxvi).

(5) Projectivity (or projectuality)

The fifth lexical specification concerns the meaning of projectivity. For Heidegger, the term projectivity (*Entworfenheit*) indicates the way a being opens up to the world in terms of its possibilities of being (Heidegger, 1962, 184, 185 / SZ, 145).[2] Inspired by Heidegger's and Vilem Flusser's work in the field of the philosophy of technology, and by Helmuth Plessner's philosophical anthropology, this book understands the concept of projectivity as the innate openness of human beings to construct themselves and their world(s) using technical artifacts. Borrowing the words of Robert Musil in *The Man Without Qualities*, projectivity is understood as "a conscious utopianism that does not shrink from reality but sees it as a project, something yet to be invented" (Musil, 1996, 11). In this projectual sense, technology can be revealed as the materialization of mankind's tendency and aspiration to overcome its thrownness (*Geworfenheit* in the original German edition of *Being and Time*) – to surpass the physical, perceptual, cognitive, and communicative limitations that define *Dasein* in its historical being-in-the-world.

Despite Heidegger's valuable and pioneering contribution to the modern field of the philosophy of technology, his work is not immediately employable in pursuing the purposes of this book. My philosophical project, as introduced in the preface, has the overall objectives of:

1. providing an ontological account of what everything looks like to our digital-hammer-wielding culture (a hermeneutical perspective), and
2. exploring the expressive limitations and possibilities inherent in using virtual worlds specifically as philosophical mediators (that is to say, as ways to materialize philosophical concepts, perspectives, and thought experiments) and as technologies of the self.

Heidegger's approach to technology is sweeping and is generally carefully framed in its relationships with philosophical thinking, but it is also inevitably limited in its applicability to factors connected to both the historical dimensions of his inquiry and to Heidegger's own overarching philosophical intentions. For example, Heidegger's philosophy of technology was developed in relation (and largely as a reaction) to progress made in the last century in machine technology and the mechanical sciences. Since he developed his philosophical project (and in particular his philosophy of technology) before the proliferation and the penetration of digital mediation in social processes and practices, it is to be expected that Heidegger's frameworks are not going to be suitable, without adaptations and re-thematizations, for mapping the specific ways in which digital technology is currently involved in cultural change, human evolution, and the development of philosophical thought.

The direct employment of Heidegger's philosophy of technology as a general framework for understanding *How to Philosophize with a Digital Hammer* is also complicated by his fundamentally transcendental and pessimistic approach to technology. For the most part, in its being monolithic and deterministic, Heidegger's philosophy of technology must be recognized as ill-suited to structuring a philosophy of openness – a way of looking at technology projectively as an integral, constitutive component of what it is like to be a human being-in-the-world. This aspect of Heidegger's thought in relation to my philosophical project will be discussed in more detail in Chapter 2, Meta-Metaphysics.

To render Heideggerian thought suitable for providing an ontological account of what everything looks like to our digital-hammer-wielding culture, this book integrates Heidegger's work with several additional perspectives. To some extent, this approach diverges from his seminal path. Instead of solely relying on Heidegger's original (existential) phenomenology, I align more closely with a position that is customarily referred to as postphenomenological. I consider postphenomenology as a set of theories capable of providing a more balanced and constructive interpretation of technologies as actively co-shaping people's being-in-the-world – their perceptions and actions, experience and existence. To clarify how I build my perspective upon a largely postphenomenological framework, a crucial starting point is to explain what postphenomenology is.

In this book postphenomenology is understood as having two interrelated meanings:

1.2.1 The praxis-oriented model of postphenomenology

As originally proposed by American philosopher of technology Don Ihde, this interpretation sees postphenomenology an approach that can offer a perspective on the philosophy of technology that can rise above the shortcomings of classical phenomenology by taking into account the context-dependence of human knowledge. Following the classical phenomenological tradition, Ihde presented the objective of technology as revealing dimensions and qualities of the actual world that could not be observed or experienced without the mediation of technical instruments. Unlike classical phenomenology, however, Ihde considers that the development of a relationship between human beings and reality precedes the theoretical establishment of a subject and an object of observation.[3] This means that, according to a postphenomenological perspective, human beings and their worlds are always mutually constitutive in their fundamental interrelation. In the constitutive encounter between humans and reality, a specific "objectivity" of reality arises (a world), as does a specific "subjectivity" of human beings (Verbeek, 2005, 130). In this sense, when trying to understand virtual worlds from a postphenomenological perspective, the role of the digital medium as mediator cannot be regarded as taking place between subject and object. Mediation needs, instead, to be understood as a way in which subject and object mutually constitute each other and, in their relationship, can be neither isolated nor absolutized.

1.2.2 Postphenomenology as the phenomenological approach to posthumanism

Posthumanism is a cultural movement that affirms the possibility and desirability of altering the way human beings are in the world through the development of technologies capable of overcoming human cognitive, biological, and operative limitations (for instance their mortality, the limited extent of their memory, the processing power of their brains, etc.). In his work, Ihde showed that, at least in contemporary Western culture, potentially all human perceptions and actions are already mediated by technical devices. From this perspective, humans can be understood as always having been "cyborgs": hybrid beings that are constituted, defined, and influenced both by their qualities as biological organisms and by the qualities of their artificial "extensions" (Haraway, 1991). According to this second meaning of postphenomenology, human beings and the worlds that they can experience are embraced as products of (technological) mediation, and not as the conceptual extremes between which mediation takes place.

In the classical phenomenological approach, technology consists of a particular – and particularly reductive – relationship with the world. Heidegger's later work in the field of the philosophy of technology is a paradigmatic example of this perspective. Heidegger presented technology not as something strictly technological (i.e., related to the material production of a certain thing or good), but rather as a particular mindset. The technological mindset is, for Heidegger, a supremely dangerous ontological "enframing" through which men seek a lording over and total mastery of nature (Tabachnick, 2006, 96). It is a reductive and derivative way of approaching the world, from which the world emerges simply as a storehouse of raw materials lying ready to be used and manipulated by human beings.

In contrast, in the postphenomenological perspectives offered by Ihde and Verbeek, the mediation of human action and perception via technological artifacts does not necessarily entail an impoverishment of the alleged ideal of a whole and authentic reality. Neither does it imply a degeneration of how human beings are "destined" (to use Heidegger's words) to perceive the world and operate in it. Rather, technologies are identified as fundamental mediators of the relationship between human beings and reality. In their mediating role, technologies are thus recognized as having the potential for disclosing new ways in which reality can manifest itself, and new possibilities for humans to shape reality and be shaped in return.

A postphenomenological approach to the philosophy of technology understands technical artifacts in terms of their capabilities and their effects as mediators – as concrete artifacts which are never "in themselves" but are always understood in relation to the human beings who engage with them (Ihde, 1990, 125). In accordance with the postphenomenological perspective, in this book technology is not conceived of as an abstract and alienating force that has the tendency to prevent humankind from experiencing the "full richness" of existence, rather, technology is embraced as the context in which a new humanism has already begun to arise.

1.3 Ontological machines?

In the digital era, Michael Heim was the first scholar who explicitly addressed the digital medium in its capability to influence worldviews and foster social change by effectively affording the experience of virtual worlds. In his 1994 book, *The Metaphysics of Virtual Reality*, Heim tried to capture what he recognized as a metaphysical shift in Western culture. This shift, Heim claimed, is caused by the capability of computers to produce effective digital alternatives to the way human beings experience and rationalize their everyday involvement with the world. Heim's original perspectives and observations were, however, presented as a fragmented collection of essay-like chapters, lacking the overall coherence and focus of a systematic model. One of the most interesting aspects of his work is his attempt to extend and re-thematize perspectives and reflections originally proposed by Heidegger (in relation to machine technology and the blossoming of mechanization), to be more directly relevant in the age of digital technology and virtual worlds.

To understand better why Heidegger's philosophy of technology might be a desirable lens through which we can attain a better understanding of the effects of the integration of the digital medium in social processes and practices, I will give a brief introduction to some fundamental aspects of his thought. A more thorough elaboration on this foundational framework is presented in Chapter 2.

Heidegger developed his perspectives on how technology shapes and influences the relationship between humans and the world during the later phase of his philosophical work. Two texts are particularly important, both derived from a 1949 series of lectures titled "Insight into What Is." The two essays of particular note are "The Question Concerning Technology" (1954) and "The Turning" (1961). A joint reading of these essays presents the world's progressively increasing involvement with technology as an element of great danger for humankind. In extreme synthesis, Heidegger identifies the threats inherent in technology as more profound and menacing than the challenges to the primary ontological role of human beings presented in the literary examples cited earlier in this chapter. The danger posed by technology should be identified, according to Heidegger, as the "coming into presence of the enframing" (Heidegger, 1982, 41–43). The "Enframing" (*Gestell*) consists of an objectifying gaze on the world, a particular declination of rationality that understands everything that exists as a resource that can be employed and exploited with a functional scope in mind. Heidegger was particularly concerned by the fact that we are not aware that we are looking at the world in that particular frame of mind, which "remains veiled and disguised. This disguising is what is most dangerous in the danger" (Heidegger, 1982, 37).

There are two main arguments for techno-pessimism that emerge from Heidegger's writings on the philosophy of technology:

- **our progressive losing sight of the dangers inherent in the technological mindset**

  The ongoing mechanization of biological, social, and productive processes fosters and reinforces the understanding of the world as a "standing-reserve" (*Bestand*): a resource whose existence is only justified by the possibility of being transformed and exploited for human purposes (Vattimo, 1991, lii; Richard Villa, 1996, 182). This orientation in relation to the world is, according to Heidegger, doubly dangerous. Apart from the evident objectification of the world inherent in understanding it as a functional system of resources, such an understanding progressively makes it more difficult for humanity to contemplate and adopt alternative worldviews. Technology is characterized by the inherent danger of becoming a totalizing perspective. When the technological mindset holds sway, Heidegger wrote, it "drives out every other possibility of revealing" (Heidegger, 1982, 27). In spite of (in fact, because of) the entire set of scientific apparatuses and theories meant to structure the knowledge of the world, humans will grow progressively less open to the "coming into being" of Being and will fail to understand deeper and more substantial truths about it (Heidegger, 1982, 27).

The second aspect of "danger" that Heidegger warns us of is:

- **the objectification of human beings themselves**

  An often quoted passage from Heidegger's "The Question Concerning Technology" states that "the essence of technology is nothing technological" (Heidegger, 1982, 35). As already discussed, technology is not presented in his work as something tangible, but rather as "a form of revealing" (Heidegger, 1982, 12, 13). The technological mindset extends the attribute of "usable object" to people, as if they were mere exploitable objects in the world. When man "is nothing but the orderer of standing-reserve, then he comes to the very brink of a precipitous fall," wrote Heidegger, "... where he himself will have to be taken as standing-reserve" (Heidegger, 1982, 27). The risk of alienation, and the social-ethical consequences of the ontological shift inherent in understanding people as an exploitable resource, deeply concerned Heidegger and guided the work of social theorists of the same period, including, notably, Karl Marx and Arnold Toynbee.

In Heidegger's thought on technology, the progressive and objectifying mechanization of the world is presented as a de-humanizing and deeply worrying prospect. In "The Memorial Address," written in 1959, Heidegger went as far as explicitly proposing an "antidote" for the technological mindset; that of treating technological products with diffidence and being ready to abandon them and live without them at any given time. Only by maintaining detachment can we preserve our critical thinking and humanity. In short, Heidegger believed that the only form of freedom that human beings can attain *in relation to* technology is freedom *from* technology.

Nevertheless, despite the techno-pessimistic aspects of his thought, Heidegger did not advocate for regressionism nor neo-Luddism. Heidegger specifically acknowledged in "The Question Concerning Technology" that it would be unreasonable to condemn and abandon technology completely. It would be utopian and naïve to hope for such radical change, as Heidegger himself recognized technology as a useful and deep-seated component of Western society.

Not every theorist who scrutinized the relationship between technologies and societies foresaw a gloomy future for the technologically-involved Western culture. Daniel Bell, for example, famously envisaged an increase in the volume and importance of information circulating in the world, driven by communication technologies. The expected quantitative increase of information would, in turn, provoke a qualitative change to society. In Bell's technologically deterministic point of view, the mechanization of the world will lead to the emergence of a "new consciousness," resulting in the emergence of a "caring society" (Bell, 1973, 15–20). Another remarkable example of a techno-optimistic approach to technology can be found in the work of Czech-Brazilian, twentieth century philosopher Vilém Flusser, who prophesied in the sixties that the technological implementation of a telematic culture would establish a relationship of mutual respect among individuals. For Flusser, technological development was to be interpreted as the continuation of the Enlightenment project begun in the eighteenth century by thinkers such as Locke, Hume, Rousseau, and Kant.[4]

The work of Vernor S Vinge, Ray Kurzweil, Hans Moravec, and of trans-humanistic currents in general, are vivid contemporary examples of strong techno-optimism. Both Enlightenment and trans-humanism approached the relationship between humanism and technology in a way that is almost diametrically opposed to the dystopian perspectives of the techno-pessimists, and yet both viewpoints can be fully ascribed to the humanistic tradition.

It is hard to detect traces of techno-optimism in Heidegger's work. However, contemporary philosophers, including David Tabachnick and Gianni Vattimo, spotted a glimmer of hope in the depths of Heidegger's techno-gloom, a solitary, poetry-inspired ray of light. Heidegger's high regard for poetry should not be surprising, as Heidegger's early thought consistently positioned the experience of poetry at the center of his philosophical reflection on truth. In Heidegger's later writings, poetry and technology are presented as two different and divergent ways of disclosing the world: in poetry, Being is understood as a form of openness (as a way of being-in-the-world projectually); whereas technology dangerously frames Being as an object and a functional, calculable resource. How could technology be understood poetically from a Heideggerian standpoint? How can technology contribute to a wider and genuine revealing of Being? How could we ever see technology as a constitutive dimension of humanism instead of a "supremely dangerous" threat to humanity?

Tackling those interrogatives in the age of digital simulations, I find it relevant to observe that Heidegger obscurely hinted at a possibile answer by citing, both in "The Question Concerning Technology" and in "The Turning," the following verses of German poet Friedrich Hölderlin:

> "But where danger is, grows
> The saving power also.
> [...]
> Poetically dwells man upon this earth."[5]

The reason why Heidegger often quoted Hölderlin's poetry when discussing the dangers inherent in the technological mindset can be identified in the similar intuitions that both authors seem to share with regard to the possibility of finding salvific potential for humanity lying dormant in the depths of mechanization. Salvation, Heidegger thought, should not be understood as a secondary aspect or a by-product of the dangers and threats with which technology challenges humanity; the danger is itself the saving power. "The danger is the saving power, inasmuch as it brings the saving power out of its – the danger's – concealed essence that is ever susceptible to turning." (Heidegger, 1982, 42)

Neither the "liberating" potential of works of art, nor the "saving power" inherent in technology was ever discussed in detail by Heidegger. Rather than giving a definitive account, he simply presented these ideas as something not only hidden and potential, but as always present in the danger *as* danger. In "The Question Concerning Technology," Heidegger prophesized that the saving power inherent to the technological mindset, "as yet inexperienced but perhaps more experienced in the future," could become a determining factor in both accessing and expanding philosophical thought (Heidegger, 1982, 33, 34).

The perspectives and ideas discussed above could be read and re-thematized in the age of digital mediation (with a degree of interpretative freedom) as an exhortation to, consciously and critically, develop, disclose, and experience philosophical ideas by means of technology. The use of technology as a means of revealing worlds and philosophical approaches could allow philosophers, and culture in general, to integrate (or even sidestep) the abstraction of descriptive language, which Heidegger often juxtaposed with poetic language. Heidegger himself clarified that, in its essence, "[*t*]*echné* belongs to bringing-forth, to *poiesis*; it is something poetic" (Heidegger, 1982, 13). David Tabachnick noted the peculiarity that, in Heidegger's thought, *techné* appears to be "both the start and the finish of, the contagion and the cure for, the most profound threat to human existence" (Tabachnick, 2006, 94).

The often overlooked philosophy of technology of the twentieth century philosopher and psychiatrist Karl Theodor Jaspers is somehow comparable with the pseudo-optimistic passages of Heidegger's summarized above. Jaspers also cautiously envisaged the potential for technology to play a role in establishing relationships with the world that are more genuine and thorough than those imposed by the blind and relentless mechanization of the world. He attributed to technology the capability of opening new perspectives, for instance by widening and deepening the perceptive abilities of human beings (Jaspers, 1951, 179, 180; Verbeek, 2005, 22). However, in early Heidegger and early Jaspers a positive and hopeful embracing of technology can only be discerned in passing remarks, while the predominant understanding of technology's effects on people is that of blinding, distancing, and alienating them from themselves and from the world with which they are natively engaged.

In the context of this book, the chance to access and activate what Heidegger calls "the saving power" inherent in (digital) technology does not suggest a voluntary return to what can be considered a less adulterated past. It is also not intended as the premise for a cultural anthropization of technology, as proposed, for example, by Gilbert Simondon or Pierre Lévy (Costa, 2007, 62). In line with the postphenomenological tradition, *Virtual Worlds as Philosophical Tools* does not embrace the concepts of freedom and genuineness as definitive for what it is like to be in the world as human beings. This is, instead, understood as a state of affairs that is always characterized and limited by biological and historical factors.

Several influential philosophical approaches have been developed from the embracing of a similar, and similarly derivative, understanding of human subjectivity. This is, for instance, the case in Ernst Kapp's philosophy of technology and Helmuth Plessner's philosophical anthropology, both of which were drawn upon for the development of the arguments presented in this book. In a way that more closely relates to media studies and media philosophy, the understanding of human beings as fundamentally lacking, limited creatures is the essential foundation of Marshall McLuhan's theoretical work. According to McLuhan, technological development can be better understood as "extensions of man"; a materialization of mankind's historical aspiration to overcome its physical, cognitive, and communicative limitations (McLuhan, 1994; McLuhan, 2008). Flusser cast light on the same idea from a different angle. Discussing the term "humanities," as opposed to "natural sciences," Flusser commented that this academic separation already presupposes an understanding of human beings as "unnatural animals" (Ströhl, 2004, 3). Plessner analogously described human beings as creatures who are "artificial by nature," open to (re)constructing themselves and their world by means of technical artifacts. This is true, following Plessner's "first anthropological law," at least from the very moment that *Homo habilis* manufactured the first stone tools (Plessner, 2006, 334).

1.4 A user's manual (chapter summaries)

In this introductory chapter, a few fundamental questions have been raised in relation to the growing integration of virtual technologies and social processes and practices. All these questions concern how the experiences of virtual worlds affect the ways in which human beings perceive, understand, and operate within their worlds. The outlined "questions concerning digital technologies" are recognized as having a fundamental cultural relevance in the digital age precisely because these questions problematize the ways in which computer-simulated worlds frame and guide our understanding of what it is like to be a human being. The following chapters will frame these questions and try to cover their effects on social processes and practices from perspectives that are informed and inspired by a philosophy of technology, media theory, and the design of videogames.

Some of the following chapters focus specifically on the methodological aspects of this philosophical project (Chapters 2 and 3 formulate the questions and present the perspectives from which the questions are asked and from which answers will be proposed). Other chapters offer ways to understand the influences that the experiences of virtual worlds have on specific social processes and practices (ranging from entertainment, to training, to experimental philosophy, to political engagement, Chapters 6 and 7). Additionally, two chapters address the question of the limitations and effects that computer-simulated worlds and experiences can have on us specifically as creators of the virtual worlds and experiences introduced above (Chapters 4 and 5).

As a tool to orient the reader within the structure of my argument, here are summaries of the chapters in relation to "the questions concerning digital technologies" introduced in this chapter. These summaries also serve to highlight and locate the key topics and claims discussed and articulated in each chapter.

Chapter 2: A Reflection on Metaphysical Thought and Its Technological "Overcoming" – To establish a philosophical foundation on which to build an understanding of technology and its effects on the perceptual, cognitive, critical, and operational capabilities of human beings, this second chapter presents a way to frame the concepts of ontology and metaphysics inspired by the work of Heidegger. Ontology is specifically understood as the rational organization of a specific set of perceptual, cognitive, and operational relationships that are constituted between a being and a world. This understanding of ontology presupposes inextricable relationships between beings and worlds that are always shaped by socio-cultural factors and contextual determinants, including biological elements. The second concept, metaphysics, indicates a family of worldviews that are typical of Western thought and that result from the establishment of a theoretical standpoint. This chapter proposes the pursuit of the fundamental cultural task of overcoming the limitations that are specifically inherent in metaphysical thought through the mediation of virtual technologies.

Chapter 3: Worlds in the Age of Digital Simulation – This chapter articulates an understanding of the effects of digital simulations on traditional human kinds of ontologies; on people's capability for structuring thought and rationalizing experience in relation to the actual world. It first establishes a definitory distinction between reality and simulation, and then focuses its attention on the capability of virtual simulations to afford effective, objective experiences of worlds. Digitally-mediated worlds are recognized in this third chapter as capable of disclosing alternative ways of understanding concepts such as time, space, and causation that are unlike the understandings through which human beings structure their everyday relationships with the world they share as biological organisms.

Chapter 4: Thinking with Virtual Worlds – As a philosopher who designs videogames and as a game designer who is passionate about philosophy, I develop an understanding of virtual worlds as philosophical tools. This chapter argues that, when presented as virtual experiences, philosophical concepts can be explored in ways that, unlike traditional forms of mediation, do not rely on subjective imagination. Through digital mediation, moreover, ideas, concepts, and thought experiments can be designed, repurposed, and manipulated, to take up a novel projective dimension. Additionally, virtual worlds can be used as technologies of the self: a means through which we, as designers, can materialize our worldviews and ideas, making them objects for our own critical evaluation and instruments that assist us in shaping ourselves and our ethos.

Chapter 5: Augmented Ontologies and a Challenge to Western Philosophy: Videogames and Simulations as Mediators of Human Thought and Experience – This chapter presents practical ways in which videogame design and the craft of building digital worlds can materialize and disclose philosophical ideas, perspectives, and thought experiments. Videogames and simulations are, accordingly, understood as mediators of human thought and human experience that can challenge and supplement the exclusively textual tradition of Western philosophy. "Doing philosophy" with virtual worlds is discussed in this chapter with supporting examples from commercial and experimental videogames. I designed and developed three videogames that were developed specifically with the objective of demonstrating and discussing the practical possibilities of augmenting and fragmenting human ontologies through virtual worlds. These videogames are the springboard for specific philosophical reflections on the possibilities and the limits of digital simulators as philosophical mediators.

Chapter 6: Positionality in the Digital Age: Virtual Bodies and the Effects of Virtual Experiences – This chapter offers an anthropological understanding of the effects of virtual experiences. Adapting Helmuth Plessner's theory of positionality for the age of computers, digital simulations are understood as capable of providing human beings with supplementary, virtual bodies. Any ontology is inevitably structured through our bodies as experiential tools. As articulated in this chapter, the experiential horizons disclosed by our new, virtual bodies offer possibilities for fragmenting, extending, and distorting human kinds of ontologies. These effects are recognized as having two possible, non-mutually-exclusive effects on our engagement in socio-political processes: they can make us more prone to envisaging and evaluating alternative states of affairs; and they can trivialize and belittle the importance of knowledge, human life, and the actual historical process.

Chapter 7: Virtual Worlds as Poetic Allegories – Inspired by the work of Paul Ricoeur, this chapter proposes an understanding of virtual worlds as poetic allegories with the scope of framing the ways in which virtual worlds disclose meaning. Anthropologically speaking, the ambiguity and the brokenness that characterize the existential condition of mankind are recognized as fundamental motivations behind the creation of technical artifacts and the development of culture, including the establishment of artificial, alternative worlds. This chapter explains that, as in the case of allegories, the effects of the experiences of virtual worlds are also deeply rooted in the specific way in which their mediator frames and presents information and worldviews. However, as is the case for any other technology, these characteristics are largely outside the complete control and understanding of the technology's creators.

Chapter 8: Virtual Worlds and the Human Condition: Cognitive, Perceptual, Critical, and Operational Limitations – Reflecting on "the questions concerning digital technology," this chapter identifies our relationships with virtual worlds (both as creators and as embodied agents within them) as inextricably tied up with and bound to the cognitive, perceptual, and operational limitations that define us as human beings. The metaphors in virtual worlds emerge from inevitably human contexts, computers as mediators are nothing but "humans who calculate," and virtual experiences are created to be encountered by human subjectivities. Consequently, questions regarding the use, effects, and possibilities of virtual technology can only be asked (as for any other technological form) as a derivation of a more fundamental question concerning the human condition. Attempting to delineate the limits of human technologies, as such, is bound to remain an open question.

2 A Reflection on Metaphysical Thought and Its Technological "Overcoming"

As established in the introductory chapter, this book offers original perspectives that are informed and inspired by the philosophy of technology, media theory, and the design of videogames. According to the general, postphenomenological approach that will be articulated from this chapter onwards, the interaction with virtual worlds (the way we experience them as beings in those worlds), as well as the very crafting of those artificial worlds (the way in which we relate to them as designers and creators), are recognized as activities capable of affecting our cognitive, critical, and operational capabilities. As such, reflecting on the experience and design of the virtual worlds of videogames and digital simulations cannot be treated as anything but a deeply philosophical task.

In order to explain the ways in which virtual worlds affect us and can be understood as philosophical artifacts, this chapter temporarily sets aside discussions concerning specific topics of philosophy of technology or media theory. Instead, it focuses more closely on the general ways of framing our relationships with technology, on the possibilities disclosed by the technical augmentation of human beings, and on an initial exploration of the effects of technical mediation on the development of thought throughout the history of Western philosophy. The objective of this chapter is to reflect on how technologies can assist and accompany us in overcoming what Heidegger called "the metaphysical heritage of thought." In pursuing this purpose, the next section clarifies what the term metaphysics indicates in the context of this book.

2.1 Meta-metaphysics

In the first century BC, the peripatetic philosopher Andronicus of Rhodes edited and arranged the writings of Aristotle. In his arrangement, he coined the term metaphysics as a bibliographic coordinate to indicate the group of books dealing with "First Philosophy," which Andronicus placed after (*μετά–meta*) those dedicated to the study of the general properties of nature (*φύσις–physis*).[1] Aristotle characterized his "First Philosophy" as the study of "being *qua* being" (the understanding of being as such, the primary and most fundamental form of knowledge) (*Metaphysics*, VI, 1026 a, 31). Although very basic, this description captured a central aspect of the traditional understanding of metaphysics, that of its aspiration toward a form of knowledge that is independent from empirical observation and universally applicable. In the Aristotelian "First Philosophy," the study of natural theology, universal science, and "being" ("meant in many ways," as specified in the seventh book of his *Metaphysics*) is explicitly performed at the level of their general and ubiquitous features.

The philosophical effort contained in Aristotle's *Metaphysics* can be read as an attempt to reconcile Plato's Theory of Forms (to which he was exposed during his years at the Academy in Athens) with the worldviews of the natural philosophy of Greece's Classical period. Interpreted in that historical perspective, it should not be surprising to observe that Aristotle undertook, in the original organization of his "First Philosophy" a broad variety of topics with very diverse levels of abstraction, from speculations on cosmogony and theology (in the books *Epsilon* and *Lambda*) to the practical study of motion and the properties of natural elements (in *Kappa*).

In the "First Philosophy" section of Aristotle's texts, for the first time in the history of written thought, one of the fundamental questions of traditional metaphysics, that of mind-dependency of knowledge, received its explicit formulation. The mind-dependency problem can be briskly formulated as: to what extent does knowledge depend on the qualities and possibilities of a human's sensory and intellectual apparatus?

A concrete example might be helpful in clarifying the meaning of these questions. Bat's meat is a delicacy customarily consumed in the Philippines and in Papua New Guinea, but it is not generally considered an appetizing ingredient in other cultural and geographical contexts, including most Western societies. This incongruence in the perception of the culinary appeal of that specific flying mammal reveals its appetizing quality not as a property inherent in the roasted bat itself, but rather as one that resides in the subjective tastes and values of the peoples relating to it as an edible item. The intrinsic subjectivity of any sensory perception is commonly experienced by humans as they engage with the world and other human beings in their everyday lives.

The very question raised by Aristotle could have been posed – and has in fact been raised – with the objective of problematizing the possibilities for humans to gather, organize, advance, and communicate knowledge since the dawning of Western thought. Radical philosophical stances in both Eastern and Western cultures have pushed this fundamental interrogative to its most extreme – and perhaps most paradoxical – consequences, and have gone as far as to cast doubt on the very existence of physical objects outside of their possibility for being perceived.[2] Because of the basic relevance of the extent to which certain basic epistemological assumptions are mind-dependent as the foundation of knowledge, disputes between realists and anti-realists are frequent throughout the history of philosophy.

Although not literally expressed in terms of mind-dependency, several approaches that are dubious about the possibility of ubiquitous and absolute qualities of reality and about our possibility to acquire knowledge about these qualities pre-dated both Plato and Aristotle by at least a few centuries. The subjectivism proposed by the early atomists in relation to the problem of induction first cast doubt on the possibility of an objective and complete understanding of the world as early as the fifth century BC. Protagoras and the sophists also openly presented knowledge as a relative construction based on a perception and interpretation of phenomena, rather than on humans accessing absolute truths. Notwithstanding the problematic state of the sources, the introduction of skepticism by Pyrrho from Elis (ca. 360 BC–ca. 270 BC) (Mancini, Marzocchi, Picinali, 1993, 169) is commonly identified as marking the transition from an approach to knowledge that relied on a general diffidence to a certain philosophical method. Attempting to condense and formalize the teachings of the philosopher of Elis in the *Pyrrhonian Writings*, Sextus Empiricus defined skepticism as "the capability to establish any kind of antitheses between phenomena and intellective perceptions" (ibid.). Given their dependence on human perception and interpretation, and thus on the limited and fallible human sensory and intellectual equipment, Pyrrho considered the achievement of any certainty regarding the correspondence between knowledge and truth to be simply and conclusively unattainable.

Since its first methodological introduction in the work of seventeenth century French philosopher René Descartes (1596–1650), any form of knowledge founded on the systematic separation of the subject and object of observation has been understood to have consequences for the very nature of the subject and the object of observation. Descartes embraced the essential stance proposed by the skeptics according to which there is no aspect of either the experience or the knowledge that humans can accumulate that cannot be encompassed by fundamental doubt.[3] The framework adopted by Descartes recognized in the outlined "methodological doubt" a necessary philosophical background for the establishment of a solid footing for the rational construction of any system of knowledge. This foundational skeptical stance was then subverted through a *reductio ad absurdum*; with his *"cogito, ergo sum"* ("I think, therefore I am"), Descartes avowed the possibility that a self-questioning mind (*res cogitans*) could exist independently from its material substrate and from any corporeal being (*res extensa*) (Descartes, 1637, Part IV).

The dichotomous structuring of the relationship between *res cogitans* and *res extensa* elaborated by Descartes gave way to a variety of philosophical perspectives (collectively referred to as dualisms) that characteristically regard mind and matter as belonging to separate ontological categories. One the one hand, Descartes identified the mind with phenomena that he considered exquisitely immaterial: consciousness, self-awareness, and the imaginative recall of mental representations (Descartes, 1637, Part IV). On the other, he understood matter as being characterized by physical dimensions (extensions) that could be encountered sensorily and to be understood through objective measurement. Through this fundamental split, Descartes' offered a philosophical approach in which one could observe and understand nature from an exclusively quantitative point of view. Plessner observed that the interpretation of the physical world upheld by Descartes, as well as its cultural success and consequent diffusion, left only two options for the development of Western thought:

> either interpreting the qualitative aspects of what exists (as well as our own bodies) mechanistically, which is to say interpreting them quantitatively, or, avoiding this analysis, explaining them as cogitations, contents and products of our interiority. (Plessner, 2006, 63, 64, my English translation from the Italian edition)

Over the last three centuries, Cartesian thought and its characteristic dualism have been subject to heavy methodological criticism from different quarters of the philosophical debate, especially in contexts such as philosophical anthropology or the philosophy of mind, where inquiries specifically concern the characteristics of experience, thought, and action, and their mutual relationships.[4] Heidegger, for example, criticized Descartes' epistemology on the basis that it established the human being as its absolute ontological center. This perspective, proposed by Plato and methodologically structured by Descartes, led all modern philosophy into a subjective stance (and thus to the objectification of truth and knowledge). According to Heidegger, Descartes disregarded the more fundamental question of what makes such a subject possible, and stunted all further ontological inquiry, bringing philosophy to today's dead end.

Regardless of Descartes' philosophically problematic dichotomy between *res cogitans* and *res extensa,* and also notwithstanding the criticism it received due to its misalignment with empirical proof, Plessner argued that not all of Descartes' epistemological efforts should be demonized or discarded. Although Descartes can be considered responsible for the elaboration and diffusion of dualistic perspectives – epistemological approaches that had a momentous methodological influence frequently considered as having been detrimental to modern and contemporary philosophy – Plessner believed that Descartes should be acknowledged for his focus on the autonomous essence of the human being and that his approach merits praise for having made possible an understanding of the human condition unconstrained by theology[5] (Plessner, 2006, 8–10; Rasini, 2010, 170). On a similar note, Hubert Dreyfus observed that among the most striking consequences of the Cartesian epistemological shift is the assumption that theoretical knowledge can be achieved in every natural domain or human activity (Dreyfus, 1991, 45).

As discussed at the beginning of this chapter, in the Aristotelian framework – and prior to the modern history of science generally – questions about the general properties and functioning of the physical world were addressed as natural philosophy. Plessner noted that, after Descartes, theoretical knowledge forced "natural philosophy" out of its original shape into a system of techniques, empirical activities, and principles of reasoning known as "the scientific method" (Plessner, 2006). Ever since, and even more systematically starting from the end of the eighteenth century, the term "metaphysics" came to denote a form of non-empirical enquiry into the nature of existence.

The successful practical applications of the empirical sciences and the wide-ranging advancements that they fostered (for example in the fields of industrial production, medicine, et cetera) contributed to the branding of ontology as an exoteric, speculative discipline, relegating it to a secondary cultural role.

It is also useful at this point to introduce, in brief, the distinction between the concepts of ontology and of metaphysics as originally set up in Heidegger's 1927 work, *Being and Time*. In the larger context of Heidegger's philosophy, the notion of ontology is presented as the structured comprehension that human beings develop about the world(s) with which they are in a mutually constitutive relationship. Ontologies, then, are natively structured by human beings both in relation to the world and in relation to themselves. Developing ontologies is a fundamental part of the characteristic ways in which human beings are in a world. Understood in this way, the notion of ontology embraces a vast and flexible horizon of possibilities in which people can understand and construct both the world and themselves "projectually."

In trying to articulate difference between the concepts of ontology and metaphysics, it is important to explain that Heidegger considered metaphysics to be a specific kind of ontology, a particular way of structuring worldviews. In his work, metaphysics indicates the general, defective ontological relationship with the world set up and fostered by Western thought. Heidegger attributed an unequivocally negative connotation to the term metaphysics, clarifying that it is a "perverted" kind of ontology, one that privileges a theoretical stance and, through its gaze, transforms everything into an object, a quantifiable resource. He associated the term metaphysics with the faulty, dichotomous epistemology that presupposes a separation between an observing subject and an observed object, a tradition of thought that originated with Plato's Theory of Forms, which obtained a methodological formulation with Descartes' dualistic epistemology and misled all philosophical pursuits within the Western tradition of thought.

From this perspective, metaphysics is identified as the fundamental and fundamentally perverted pursuit of truth as an objective property. This "perversion," Heidegger argued, is largely overlooked in the methods and goals of Western thought and it is hardly ever questioned or problematized. According to Heidegger, it is metaphysical thinking that led philosophy up the wrong path to the extent that it reached the paradoxical point of questioning the very existence of reality.

As part of Heidegger's overall philosophical project, metaphysics is understood as a defective heritage of thought – a distortion, an incorporation, and a remembrance – that must be "overcome." To assess if and how a form of overcoming traditional ontological structures can be pursued with the assistance of digital technology, the next logical step is to attempt to establish the boundaries and the epistemological possibilities of pre-digital human kinds of ontologies.

To this point the term metaphysics has been assigned several different meanings. As in the history of Western thought, it has been used to indicate:

- a bibliographical reference
- the knowledge of what extends beyond the physical world (see Footnote 1)
- a derogatory categorization for certain intellectual efforts in their attempt to address purely speculative (and often extravagant) questions
- the perverted and defective framework of Western thought that philosophy should constantly problematize and overcome.

In his 2006 book, *What is this thing called Metaphysics?*, Brian Garrett proposed yet another perspective. According to Garrett, metaphysics can be better understood not as a single notion, but as the involvement of two connected concepts:

- The first concept is that of a metaphysics (with the indefinite article and a lower case m), defined as a fundamental elaboration on information deriving from human sensory experience. A metaphysics constitutes, in this sense, the primary structure for the development of any system of thought. Similar to the notion of ontology introduced in the first chapter, Garrett's understanding of "a metaphysics" denotes the basic ideas and relationships that determine how an individual or a culture approaches and shapes a specific understanding of the world.
- The second concept is that of Metaphysics (without an article and with a capital M). According to Garrett, this more general term indicates the eclectic bundle of disciplines and methods that are used in developing "a metaphysics" (Garrett, 2006, xiii).

The list of possible interpretations of the term continues to grow. To cite a more recent interpretation, Quentin Meillassoux treated metaphysics as a particular form of ontology, namely a form of ontology that proposes a dogmatic (and often religious) understanding of the very existence of a reality outside the human ability to perceive it (Meillassoux, 2011, chapter 1).

In sum, the term metaphysics has been employed over the course of the history of Western thought to indicate a philosophical field, its component disciplines, various forms of speculative thought, and an editorial convention (in a sort of multifarious, concomitant, self-referential, and contradictory Borgesian classification). The fragmentation and the ambiguity of the term demand a methodological clarification of how the term metaphysics is employed in this text.

Once again, Heidegger's work is the weapon of choice to accomplish this task, meaning that I align the use of the term metaphysics in this book to Heidegger's understanding. This decision is not only dictated by convenience (as most of the foundational lexicon used to discuss digitally-augmented ontologies in this book is already either Heidegger's or elaborated upon an Heideggerian perspective), but is also motivated by the fact that the overall argument presented in this book draws inspiration and originally builds upon two interconnected aspects of Heidegger's philosophy:

- his seminal work in the field of philosophy of technology (introduced in its relevance to my questions in the first chapter), and
- his overarching philosophical project of overcoming the derivative and unsatisfactory tradition of Western thought that he indicates with the term metaphysics.

The purpose of the following sections is to establish how such overcoming might take place, starting by defining an understanding of the boundaries of human kinds of ontologies and then trying to envisage overcoming such limitations in ways that might help us recognize and discard our metaphysical tradition.

2.2 What are the epistemological limits of traditional ontologies?

In his 1641 *Meditationes de Prima Philosophia* (*Meditations on First Philosophy*), Descartes expanded on the philosophical system originally outlined in his 1637 *Discours de la méthode pour bien conduire sa raison, et chercher la vérité dans les sciences* (*Discourse on the Method*). In the *Meditationes*, he proposed the assumption that, to perceive, understand, and in general to relate to a world, a mind must initially contain some form of internal structure (Descartes, 1641). German mathematician and philosopher Gottfried Wilhelm Leibnitz (1646–1716) also founded his dualist interpretation of knowledge on the presumed existence of an innate epistemological framework. For Leibnitz, the human soul is a microcosm, a monad, naturally endowed with an internal *vis repraesentativa* (representational power) that determines the soul's capability to perceive and interpret an external world.

The belief that an inborn cognitive structure is a functional necessity at the basis of any form of knowledge, commonly referred to as innatism, was first recorded in the history of written thought by Plato in the Socratic Dialogue *Meno*. The innatists' standpoint in the philosophy of mind is in diametrical opposition to the epistemological hypothesis known as *tabula rasa* (blank slate), which made its first appearance in Aristotle's work.[6] According to the theory of the blank slate, all beings are born without any mental content. In this vision, any form of knowledge can be understood as a construct that proceeds purely from experience. These two competing approaches constituted the premises for the eighteenth century epistemological dispute between continental rationalists, who understood knowledge as a rigorous, deductive, intellectual *a priori* construction, and the British empiricists, who believed that cognitive processes were *a posteriori* approximations based on sensory evidence.

According to the empiricist philosopher John Locke (1632–1704), the stability and the objectiveness of knowledge that can be deductively structured in relation to geometric and mathematical concepts is not achievable with regard to how human beings experience the world. Given the sensory – and thus imperfect and accidental – foundations on which human understanding of the world is based, Locke maintained that science cannot be anything other than a generalization of empirical observations and will, consequently, never be able to demonstrate any necessary relationships between physical phenomena.

In the eighteenth century, David Hume adopted Locke's conclusions concerning the nature of empirical knowledge and its necessarily practical orientation. One of Hume's foundational philosophical efforts, building upon Locke's theoretical framework, is of particular relevance when discussing human ontologies: his inquisitive exploration of the concept of causation. The starting point for Hume's inquiry on causation can be paraphrased in the following interrogative: Why is it instinctive to expect that everything that happens in the world has a cause? Hume's skepticism with regard to the realism of the idea of causation and his rejection of the unquestioned acceptance of the traditional understanding of the relationship between cause and effect had a decisive influence on the work of Immanuel Kant. In the preface of his 1793 text, *Prolegomena to Any Future Metaphysics*, Kant explained that it was the reading of David Hume that interrupted his "dogmatic slumber" and gave his investigations in the field of speculative philosophy a new perspective. In the same text, he specifically identified Hume's questioning of the traditional understanding of causation as the original inspiration for the articulation of his 1787 *Critique of Pure Reason* (Kant, 1783, 4: 257, 4: 261; Kant, 2000, 23).

For Hume, causation – as a relationship developed over time among physical phenomena – was neither intuitively evident nor logically demonstrable and, as such, could not offer any truths independent from the observation of the phenomena in question. According to Kant's perspective, Hume, albeit accurate in his inference, failed to follow his thinking to its ultimate consequences. Kant believed that Hume stopped short of considering that synthetic judgments (ways of thinking that are characterized by the fact that the predicate is not already thought in the subject) can be made *a priori* of experience (Kant, 2000, 146–148).

In contrast to Hume, Kant observed that the core of the matter was not whether the concept of causation was in itself legitimate, applicable, and indispensable in relation to human knowledge of the physical world. What Kant attempted to understand in his speculative project was whether causation could be thought *a priori* of experience. In that case, he would be able to establish the autonomy of the principle of causation from any object of sensory perception and, consequently, its intrinsic and universal validity. His philosophical effort in relation to causation in the *Critique of Pure Reason* played the role of a first, exemplary step toward the ultimate purpose of being able to determine whether the very conditions at the basis of human knowledge could be established *a priori* as logically necessary and independent from experience. Clearly for Kant, this question came bundled with an aspiration to explore the extent to which such conditions guarantee the absoluteness and universality of any theoretical postulation.

In his attempt to resolve the historical dispute between the competing epistemological approaches of empiricism and continental rationalism, Kant elaborated a philosophical framework wherein human subjects are seen as having the possibility of accessing *a priori* truths about reality. Such truths, he believed, could be achieved through analyzing the sensory pre-conditions of human experience itself. In his reflections on the foundations of knowledge and in accordance with the interpretation of space and time proposed a few years earlier by Swiss mathematician Leonard Euler, Kant understood space and time as the necessary (and, as such, given *a priori* of any experience) framework for phenomena to be perceived by humans.

What were traditionally recognized as properties of reality – such as the spatial extension of physical objects, the passage of time, and the causal relationships between events – were identified by Kant as qualities imposed on reality by the intellectual equipment of human beings. A metaphor that is often colloquially employed to encapsulate Kant's understanding of the cognitive influences deriving from our perceptual and intellectual systems depicts human beings as wearing tinted spectacles through which they observe the world. As a consequence of these metaphorical spectacles, humans are bound to experience a world with hues and deformations that are not inherent properties of reality, but that depend on the qualities of the lenses they are wearing.[7] Concordantly, in Kant's view, when synthetic *a priori* knowledge is achieved, it is actually absolute knowledge about the "tinted spectacles", the very forms of human receptivity (Kant, 2000, 192).

Whereas empiricists understood the mind as being a largely passive receptor of phenomena from the world, Kant maintained that experience is actively constructed by the mind in accordance with content-less, innate rules. Instead of interpreting the outside world as being passively and objectively received by the human mind, Kant assigned the mind the role of active constructor of human worldviews. This shift in understanding the relationship between mind and reality (or between subject and object of observation) is what Kant himself referred to as his "Copernican revolution" of thought (Kant, 1783, 4: 257, 4: 261; Kant, 2000, 23).

As a consequence of what was observed, Kant considered the *noumenal* essence of reality beyond the human beings' ability to grasp and articulate knowledge. Reality, for what it is (being *qua* being), could not be accessed by the human intellect. The *noumenon*[8] is, for Kant, the limit of pure reason. This epistemological boundary, as Virgilio Aquino Rivas pointed out, negates the pre-conditions of transcendence and rules out any aspirations to absolute knowledge (Rivas, 2007, 65). From the theoretical framework of Kant's *Critique of Pure Reason*, thus, some of the most abstract questions traditionally ascribed to metaphysical thought (e.g., Is there a God? Do humans have free will? Is there a world outside of any consciousness?) are entirely unanswerable.

A fundamental methodological aspect of Kant's perspective consists of the reduction of being human to a purely intellectual subjectivity. In the *Critique of Pure Reason*, human beings are presented in a highly abstract form, a form that – in its theoretical simplification – ignores any peculiarities or differences in how individuals relate to reality. According to Kant's transcendental idealism, all of humankind shares the same perceptual and intellectual limitations. Consequently, for Kant, the frontiers of human thought coincide with the boundaries of what he understood as the common backdrop of mankind's sensory and intellectual equipment.

A generic and abstract understanding of subjectivity is not a unique trait of Kant's work, but was shared by several other thinkers of the same period. Most of the philosophical approaches developed during the Enlightenment era relied on an understanding of the world as static and homogenous, a (reductive) interpretation that could rationally – and hence indisputably – be described in terms of the laws of nature. In that respect, even a skeptic like Hume expressed the belief that "[m]ankind are so much the same, in all times and places, that history informs us of nothing new or strange in this particular. Its chief use is only to discover the constant and universal principles of human nature ..." (Hume, 1748, section VIII: On Liberty and Necessity, Part I).

It is relevant to the development of my argument to underscore that, in the realist epistemological framework developed by Kant, reality was fundamentally perceived in the same way by all human beings. His perspectives relied on the anthropological premise that all of mankind shares identical perceptual and cognitive equipment regardless of individual, social, or historical characteristics such as, for example, religion, language, historical period, level of technological development, or literacy. Consequently, all humans can grasp and shape the material world through the same *a priori* forms of understanding. Were distinct human individuals endowed with different *a priori* forms, they would literally live in different worlds. In other words, Kant believed that the sensory projections that mankind can experience are indeed constitutive for the phenomenal world of each human subject but are, at the same time, non-subjective. In Kant's perspective, human sensory equipment is the fundamental boundary to their capability for producing, organizing, and communicating thought. In his view, humans cannot logically be considered capable of obtaining access to any truth or knowledge that are absolute (i.e. that do not depend on sensory experience).

Departing from the Kantian interpretation, a human subject's perceptions of the world and rational elaboration of such perceptions could be understood as not necessarily homogenous with those of other human subjects on the grounds that such perceptions and rationalizations are always influenced by contextual factors and individual diversions from an idealized norm. Such diversions could, for example, be blindness, color blindness, deafness, a particularly high or low intelligence quotient, attained level of formal education, type of religious belief, or sexual orientation.

Heidegger openly praised and highly valued the depth and the influence of Kantian thought, but he deemed Kant's perspective on thought not to be as revolutionary as Kant himself claimed and as too conventionally constrained within the tradition of Western philosophy. According to Heidegger, "Kant took over Descartes' position quite dogmatically notwithstanding all the essential respects in which he had gone beyond him" (Heidegger, 1962, 45 / SZ, 24). This particular aspect of Kant's work has often been the subject of academic criticism on the basis of its reductionism and its antihistoricism.

A level of abstraction in embracing the essential qualities of human beings reminiscent of Kant's conceptualization can, however, be observed in the first phase of Heidegger's thought. Despite Heidegger's acknowledgment in his 1924 book, *Der Begriff der Zeit* (*The Concept of Time*), that questions about the human being require the analysis of the "ontological characteristics" of *Dasein* in its historicity, the early Heidegger problematized "being-in-the-world" as humans in a way that is essentially ahistorical (Heidegger,1962; Heidegger, 2011, 73). In *Being and Time, Dasein* is understood from several different perspectives (e.g. through the lenses of temporality, mortality, thrownness, care, etc.), but it is generally embraced as a way of being that is experienced in the same way by all of humanity over the entire course of human history. Only after the publication of *Being and Time,* did the historical dimension of being return as a central dimension of his thought (Thompson, 2005, 114).

The belief that the possibility of developing human thought is dependent on the perceptual and intellectual limitations of mankind and is, therefore, necessarily historical, was first discussed, in a quasi-philosophical fashion, by Xenophanes of Colophon (570–480 BC) as early as the sixth century BC.[9] As a philosophical notion, it received its first elaboration in Hume's 1748 "An Enquiry Concerning Human Understanding." When discussing human imaginative capabilities, Hume observed that:

> Nothing, at first view, may seem more unbounded than the thought of man, which not only escapes all human power and authority, but is not even restrained within the limits of nature and reality. To form monsters, and join incongruous shapes and appearances, costs the imagination no more trouble than to conceive the most natural and familiar objects.... But though our thought seems to possess this unbounded liberty, we shall find, upon a nearer examination, that it is really confined within very narrow limits, and that all this creative power of the mind amounts to no more than the faculty of compounding, transposing, augmenting, or diminishing

> the materials afforded us by the senses and experience. (Hume, 1748, section II: Of the Origin of Ideas)

The idea that being-in-the-world as humans is a fundamental imprinting that frames and limits our ability to develop knowledge is a recurring philosophical theme in the continental tradition. It was famously associated with the concept of metaphysical "destiny" in the philosophical work of Georg Hegel, Friedrich Nietzsche and in various currents of the phenomenological school (Heidegger, 1962, 82 / SZ, 56; Vattimo, 1991, xlvii-liii).

With the objective of developing a new perspective on digital mediation that is capable of not only dialoguing with the philosophy of technology, but that also confronts the anthropological consequences of technological development (and of the integration of digital technologies in social processes and practices), I propose a supplementation to the argument summarized above. I believe it would be, in fact, more accurate and constructive to understand human beings as being "stuck" with the experience of what it is like to be human beings *in specific socio-cultural (and technological) contexts*. Recognizing and giving relevance to the contextual and historical qualities of human ontologies (purposefully used in its plural form) is something that this book pursues not only for the sake of completeness or methodological rigor, but also as a foundational aspect of the very purpose of this enquiry, that of understanding how technologies capable of giving rise to virtual worlds can affect us cognitively and epistemologically.

2.3 What does it mean to "overcome" traditional ontologies through technology?

In "Plato's Doctrine of Truth," Heidegger's 1942 essay focusing on the Platonic allegory of the cave, he highlighted an essential divergence concerning the question of metaphysical truth between Plato's perspectives and those of the pre-Socratic thinkers (Heidegger, 1998, 155–182). According to Heidegger, the question of being and the problematization of its meaning invested and constituted the radical objective of philosophy from Parmenides to Aristotle. The pre-Socratic philosophers conceived truth as a form of undisclosedness: the revelation of being for what it means in the most basic sense.

In his essay, Heidegger explicitly named Plato as being responsible for setting Western thought on the perverted path toward the objectification of truth and knowledge. "While for the pre-Socratics 'being' still meant 'emerging out of concealment into unconcealment,' for Plato it began to mean 'essence'" (Verbeek, 2005, 51). As already discussed, Heidegger believed that, from Plato onwards, philosophy had been perversely treated as a form of physical investigation and its objective had been understood as that of determining the meaning of being, as if beings were mere, detached objects (Heidegger, 1998, 155–182; Volpi, 1998, 91–94).

In his 1986 book *The View from Nowhere,* American philosopher Thomas Nagel identified the attitude outlined above with the term scientism. Offering a critical attitude toward the Western tradition of thought comparable with the criticism found in Heidegger's work, Nagel argued that:

> Philosophy is ... infected by a broader tendency of contemporary intellectual life: scientism. Scientism is actually a special form of idealism, for it puts one type of human understanding in charge of the universe and what can be said about it. At its most myopic it assumes that everything there is must be understandable by the employment of scientific theories like those we have developed to date – physics and evolutionary biology are the current paradigms – as if the present age were not just one in the series. (Nagel, 1989, 9)

In Heidegger's early work, the abstraction and the functional orientation of the metaphysical tradition of thought are presented in sharp dissonance with the phenomenological analysis of existence. In the first phase of his thought, and in *Being and Time* in particular, Heidegger envisaged a new beginning for Western philosophy, a new epoch of thought that would emerge from the overcoming and the forgetting of Plato's understanding of truth (Heidegger, 1998, 155–182). He overtly advocated the deconstruction and the abandonment of metaphysics as the philosophical antidote to the current and totalizing form of thought uniquely preoccupied with knowledge-theory. Heidegger's critical stance toward Western metaphysics was also frequently coupled with the position that philosophy could be efficiently "restarted" by recovering a more original form of thinking founded on the practically involved and more thorough relationship with philosophical truth that was established by pre-Socratic thinkers.

As has already been explained, Heidegger understood the concept of ontology as "the essential ground on which humans relate to the world and in which knowledge is rooted" (Heidegger, 2008, 94) and philosophy as "the painstaking effort to think through still more primally what was primally thought" (Heidegger, 1982, 22). In line with these premises, he believed in the necessary philosophical commitment to the constant questioning and destruction of traditions of thought and the intellectual imperative to constantly attempt the overcoming of their boundaries.

In constructing his radical criticism to Western thought, Heidegger explicitly stated that he was not opposed to scientism *per se*. Heidegger did not imply that Western metaphysical tradition of 'subjects opposed to objects' had no legitimacy. On the contrary, he believed that the world can be fruitfully interpreted as a set of external, interconnected objects, and that such objects can conveniently be seen as resources: as assets that can be approached and utilized for a functional purpose. In this sense, the theoretical separation between subjects and objects is precisely what made the natural sciences possible. What Heidegger was arguing for, instead, is that we do not absolutize that particular "world picture". According to him, we should rather understand that (metaphysical) stance as one among many other possible ways in which humans can relate to reality – a specific and specifically reductive way of thinking that emerged in a certain context and a particular moment in history.

Without sharing Heidegger's hope for a nostalgic return to an idealized reconstruction of the Greek origins of Western thought, I sympathize with his overarching philosophical project of critically reflecting on the thought-horizon of Western metaphysics with the objective of overcoming its limitations and pursuing a more thorough and encompassing reflection on being. One dimension of Heidegger's early philosophy that is at odds with the philosophical aspirations of this book is the ahistorical nature of his analysis of *Dasein* in *Being and Time*. This aspect characterizes Heidegger's early work. This ahistorical position prevented Heidegger's early thought from structuring itself as a philosophy of openness. Technical, biological, or political changes, or sociocultural transformations in general, cannot be understood, in *Being and Time*, as factors capable of influencing the ways human beings are in the world. In other words, cultural shifts and technological advancements are understood by the early Heidegger as exclusively the product of an epochal change in the ways humans relate to the world and not as contributing to those transformations as causes or concomitant factors. I find this aspect of Heidegger's early work to be insufficiently capable of describing the dynamism and the increasingly influential interconnections between human beings and technically-mediated social practices.

Taking a closer look at the relationships between technology and being-in-the-world as articulated in the early phases of Heidegger's thought, these can be understood as structured in a privative fashion, meaning not as novel possibilities for human beings to establish additional ways of engaging with the world, but rather as forms of severance from it, as deficient modes of being. In relation to the aspirations to openness inherent in my philosophical project, the second phase of Heidegger's thought is – at least in a general sense – more promising than the first in terms of its less monolithic approach to the projectual possibilities of our technically-mediated being-in-the-world. Although it continued to explicitly pursue the same general objectives, Heidegger's later thought demonstrates on several occasions a degree of openness toward certain historical dimensions of *Dasein*. This is particularly evident, in my perception, in his writings concerning the sociocultural role of works of art or social production in general ("The Origin of the Work of Art," "Building, Dwelling, Thinking").

Despite the uncompromising and deterministic stance adopted by the later Heidegger in relation to art and technology, several dimensions of that phase of his thought had a formative influence on the structuring of my arguments. Two of the most relevant contributions to the development of my argument of Heidegger's philosophy after the alleged "turn" in his thought can be identified in:

- the recognition that the relationships that human beings can establish with artworks and with artifacts in general have influences and effects that extend beyond their functional use. According to Heidegger, these relationships can establish new worlds and facilitate the emergence of alternative worldviews. Activities pursued with the use of artifacts, such as sailing, hunting, or erecting buildings, are, in the later phase of Heidegger's thought, understood as forms of revealing and as "scenes of disclosure" (Heidegger, 1982, 12, 13; Heidegger, 2000, 174).
- the recognition of possibilities for overcoming Western thought as latent in the depth of the technological mindset. In "The Question Concerning Technology," "The Turning," and "The Age of the World Picture," Heidegger obscurely hinted at a growing opportunity for what he calls "salvation" lying at the very core of technology.

Not all the developments in Heidegger's later thought can be, however, viewed as contributing to the cause of a postphenomenological understanding of virtual worlds as mediators of human thought and factors of sociocultural change. The most worrying among these developments is, for me, Heidegger's problematic relationship with the discipline of anthropology. I find that aspect of his thought to be particularly troublesome for my work, as the concluding sections of the arguments presented in this book (specifically in Chapters 6 and 7) rely on perspectives that are either inspired or directly borrowed by anthropology and philosophical anthropology. In the next paragraph I will outline why the attempt to combine a Heidegger-inspired perspective with notions and approaches originating in the domain of anthropology is even more delicate and complicated in the case of Heidegger's later work.

In the early phases of his thought, Heidegger approached philosophical anthropology in a rather neutral way, considering it to be the name for a "philosophical interpretation of man which explains and evaluates whatever is, in its entirety, from the standpoint of man and in relation to man" (Heidegger, 1977, Vol. 5, 86). In that period, he demonstrated a certain appreciation of such perspectives. After the publication of *Being and Time*, instead, Heidegger openly radicalized his attitude toward philosophical anthropology, which ultimately led to its philosophical rejection. In the later phases of his thought, he deemed philosophical anthropology to offer unsatisfactory perspectives that tried to understand the human being starting from its "animality". As such, Heidegger considered it to be hopeless as far as the pursuit of an original and thorough "questioning" of being (Heidegger, 1982, 140). Towards the later phase of his thought, he identified the anthropological position (understanding the human subject as the unquestionable foundation of knowledge) as an accomplished manifestation of the Western metaphysics that he so thoroughly criticized.

Heidegger's dismissal of philosophical anthropology is a very interesting dimension of his later thought that can be associated with the advocacy for a wider and deeper understanding and pursuit of what humanism means and can mean, a position that is epitomized in his "Letter on Humanism." However, it is evident that Heidegger's later rejection of what he understood as the limited and derivative horizon of philosophical anthropology constitutes a conceptual obstacle when trying to integrate this approach with his pioneering insights in the field of the philosophy of technology. In this book, I surmount this incongruence by adopting, in this particular case, the less inflexible and more fruitful instrumental understanding of philosophical anthropology that is characteristic of the early developments of Heidegger's work, rather than his later interpretations.

I would like to point the reader to the fact that there is a common denominator underlying the decisions as to which aspects of Heidegger's philosophy were embraced as constituent elements of my arguments. This common denominator is the desire to rediscover aspects of Heidegger's thought that do not reject practices and "regional ontologies"[10] wholesale as uselessly derivative or as openly detrimental. With the goal of understanding digital technologies as technologies that promote a specific kind of human openness toward modality and projectivity, it is not only logical but also desirable for the philosophical objectives of this book to re-thematize and repurpose certain dimensions of Heidegger's heritage that can be interpreted as pointing in the same direction. This philosophical purpose is pursued on the basis of the interpretation, shared by Heidegger himself, that Heidegger's work, before and after the alleged "turn" in his thought, can be interpreted not as two separate philosophical frameworks at odds with each other, but as two moments in the same process – as two aspects of a consistent philosophical project that have different focal points but that ultimately pursue the same goals.

From the outlined use of Heidegger's work, it should be evident that overcoming traditional kinds of human ontologies through the use of technologies cannot be understood as a complete and definitive abandonment of our biological and philosophical heritage. It should also be apparent that I am not advocating a dogmatic and unquestioned acceptance of the integration of digital technologies in social processes and practices. Instead, I am proposing that the fragmentation, extension, and distortion of human kinds of ontologies through the mediation of virtual worlds are always accompanied by a reflective, open, and critical attitude toward the larger contexts in which mankind develops thought, and establishes social practices and relationships.

2.4 Preliminary conclusions

In Heidegger's framework, the fundamental question on which philosophy needs to be constructed (or rather reconstructed) is deeper and more encompassing than the deficient, objectifying one passed down from Plato. In his early writings, heavily influenced by the work of his mentor Edmund Husserl, Heidegger laid the speculative groundwork for his envisaged overcoming of Western metaphysics and promoted it as a partial and faulty tradition of thought. According to Heidegger's perspective, human beings are thrown into the world with a human kind of biology and, as a consequence, are destined to structure specific kinds of relationships with reality. Given these premises, the only context where an overcoming of the objectifying and theoretical kinds of ontologies grouped under the name of metaphysics can come about – if at all – is the very context of human kinds of understanding. Consequently, for Heidegger, the concept of overcoming is not presented as a radical break with traditional ontologies, to which we are inescapably bound, but rather as their acceptance-distortion (or *Verwindung*, as explained in Chapter 1).

Inspired by the philosophical perspectives of the early phase of Heidegger's thought, this inquiry shares the belief that there is a broader and more encompassing philosophical horizon to be embraced by human beings through human kinds of ontologies. However, in *Being and Time,* the human perceptual and intellectual equipment is understood in an ahistorical fashion, as an absolute and context-independent background for knowledge shared by all of mankind. From this perspective, the *Verwindung* of people's ontological horizon could only be construed as a utopian aspiration or a remote hope. In contrast to the early developments in Heidegger's thought, this book acknowledges the key role played by sociocultural determinants in structuring and modifying human kinds of ontologies. These determinants include, but are not limited to, biological differences and psychosocial determinants, such as stage of technological development and religious orientation. In other words, to better understand our relationship with digital technology and virtual worlds this book accepts a historical interpretation of the human being. In that respect, De Mul noted that "[t]he development of hardware and software is taking place so rapidly that the whole sector seems to find itself in a permanent beta state. When we also consider that information technologies radically reconfigure almost every aspect of our society and our lives we realize how confusing our situation is. We, too, are living in a permanent beta state" (De Mul, 2010, 41).

From the perspectives outlined above

1. the relationship with technology can be embraced as an essential aspect of being-in-the-world as human beings, instead of as a derivative mindset that constrains and perverts human thought and experience, and
2. the overcoming of traditional ontologies can be projectually pursued, instead of being vaguely prophesized.

In his later reflections on technology (in "Building, Dwelling, Thinking," "The Turning," and "The Question Concerning Technology") Heidegger himself suggested various ways that the transcendence of the metaphysical horizon of Western thought beyond its traditional boundaries could take place through the apex of Western metaphysics itself, technology:

> when we consider the essence of technology we experience enframing as a destining of revealing. In this way we are already sojourning within the free space of destining, a destining that in no way confines us to a stultified compulsion to push on blindly with technology or,

> what comes to the same, to rebel helplessly against it and curse it as the work of the devil. Quite to the contrary, when we once open ourselves expressly to the essence of technology we find ourselves unexpectedly taken into a freeing claim. (Heidegger, 1982, 25, 26)

On these fundamental assumptions, I develop an original perspective in relation to the philosophy of technology, according to which the specific contribution of the interactive experience of virtual worlds in the shaping of human thought is not a revolution or a radical break with our cognitive and philosophical past, but it amounts to a deepening and a fragmentation of the possibilities for humans to perceive and understand worlds, and to operate actively within the interactive horizons they disclose.

3
Worlds in the Age of Digital Simulation

Stemming from a traditional ontological divide, the field of digital media studies still distinguishes the digital experiences accessible via computers into two broad categories:

- The first is that of telepresence, a family of technologies that affords various degrees of agency and the exchange of information between users and actual environments that are not immediately present for them. The experiences and worlds made possible by teletechnologies and robotics are, therefore, qualified as telepresent. In other words, telepresence technology allows humans to establish aesthetic and interactive relationships with their world in ways that transcend their scale, their spatial location, and, often, their native biological capabilities. The Mars Exploration Rover that was sent to Mars by NASA in 2003 is an extreme example of the experiential extension afforded by telepresence. The rover extends and dislocates the cognitive and interactive capabilities of the NASA scientists both in space (they can observe and analyze chemical samples in a part of the universe that is several million miles away), and in time (the radio signal-based interactions between NASA and the rover have an average time delay of 20 minutes between inputting the controls for action and the perception of the results).
- The second category, simulations, shares the same basic characteristics of telepresence in terms of surmounting the cognitive and interactive limitations inherent to being-in-the-world as human beings. However, instead of granting access to experiences that are not present, simulations disclose experiences that are perceptually effective, but that do not exist in the world commonly labelled as actual. From a strictly ontological standpoint, the qualities of

> simulated worlds have no necessary relation to the world humans inhabit as biological creatures, although simulated worlds are designed through (and mostly for[1]) human kinds of worldviews. Digital simulations are frequently used as training technologies, for example in the aviation or medical fields, that allow people to gain practical experience without the risk of causing actual damage to people or equipment. Besides the (often interactive) enactment of circumstances that could occur in the world indexed as actual, digital simulations are capable of providing an experience of virtual, alternative worlds and therefore have the potential to emancipate humans from the pre-digital constraint of having to shape their existences and ontologies exclusively in relation to a single world (the one we share as biological creatures, previously indexed as actual). Digitally mediated simulations offer a multiplicity of artificial experiences that are often physically dangerous or simply not possible in the actual world.

Before embarking on an exploration of the effects of an experience of simulated worlds on human kinds of ontologies, I find it important to observe – as Espen Aarseth illustrated in his seminal 1997 book, *Cybertext: Perspectives on Ergodic Literature* – that the rise of the concept of simulation in contemporary culture is, at least from a qualitative point of view, independent from the introduction and diffusion of the digital platform in social practices. Aarseth noted that simulations, and other interactive and combinatorial forms of expression and representation, made their cultural debut in the familiar form of tabletop and card games, and even in certain uses of text many centuries before the advent of the digital medium. With regard to textual examples, it might suffice to think about examples of ergodic literature[2] (such as the *Choose Your Own Adventure* gamebooks series) or the experiments of *OuLiPo*[3] at the beginning of the twentieth century (see Figure 3.1). From a comparable perspective, game studies scholar Gonzalo Frasca observed that:

> Simulation is not a new tool. It has always been present through such common things as toys and games but also through scientific models or cybertexts like the *I-Ching*. However, the potential of simulation has been somehow limited because of a technological problem: it is extremely complicated to model complex systems through cogwheels. Naturally, the invention of the computer changed this situation. (Frasca, 2003, 2)

Figure 3.1 The editorial structure of Raymond Queneau's 1961 book *Cent Mille Milliards de Poèmes* (*A Hundred Thousand Billion Poems*; drawing by Alessandra Mazzucchelli, used with permission). In the preface to the book, Queneau wrote that "This small book allows everybody to freely compose a hundred thousand billion poems, all regular of course. It is essentially a machine to produce poems ...." *OuLiPo* (see Footnote 1) began when Raymond Queneau, stalled in the composition of this book, solicited the help of mathematician Francois Le Lionnais

A similar attitude toward cultural production and cultural consumption could be observed in the diffusion, at the beginning of the last century, of installations and combinatorial works, including Surrealist collages, Dada's assemblages, and the Bauhaus' early examples of kinetic art. Such creations flagrantly challenged traditional representational canons in their multimediality and occasional interactivity, and were programmatically meant to encourage a less univocal and prescriptive approach to cultural production and to individual expression. The emergence of philosophical currents like hermeneutics and phenomenology in the same period reinforces the interpretation that all these cultural factors did not coincidentally co-occur, but rather were the phenotypes of a wider and more encompassing cultural shift that challenged the pretended universality and stability of classical worldviews.

Differently from their pre-digital mediations, computer simulations can not only manage systems that are complex and (to some degree) autonomous, but can also disclose virtual worlds that are growing progressively more detailed and immersive with the advancements of virtual technologies in terms of both computational power and the capability of input and feedback devices. Following from the phenomenological definition of world adopted in this book, the virtual worlds that can be experienced through the mediation of computers must also effectively be considered as worlds.

In setting up an understanding of digitally mediated simulations that is capable of mapping the cultural consequences of their diffusion on the background of their technological affordances, it is relevant to introduce the concept of "technological momentum," proposed in 1994 by historian of technology Thomas P Hughes. Hughes' attribution of a social inertia (or momentum) to technological systems is motivated, in his theoretical standpoint, by the need to overcome the dichotomy between the two foremost, competing models used to describe the relationships between technological development and social change. Specifically, Hughes' theory of technological momentum creates a synthesis between the largely divergent positions of technological determinism and social constructivism.[4]

In his theoretical work, Hughes reconciled technological determinism and social constructivism by recognizing the two theoretical standpoints as two ways to approach technology in the development of the relations between a certain technological system and a social group adopting it. In other words, Hughes sees technological determinism and social constructivism as two attitudes characterizing two different periods in the adoption and integration of a technology in a certain sociocultural context. According to Hughes, upon the introduction to society of a new artificial system (or a new use of an established technology), the society exerts a deliberate guidance over the technology's qualities and ways it is employed. In that initial phase, habits, canons, and social policies exercise a degree of control over how sociotechnological relationships are established. With the use of a new technology (or the novel use of an existing technology) becoming more common and ingrained in the fabric of society (thus reaching a state of maturity), the initial socially constructivist approach transitions into a more inertial and less flexible relationship with that technology that is recognizable as a form of technological determinism (Hughes, 1994). It is crucial to note that from Hughes's standpoint, a technology can be identified as both the cause and the effect of social change. Put somewhat more simply, in the processes involved in their development and integration in society, technologies are both shaping society and being shaped by society in return.

As has already been described on several occasions, the understanding of technology presented in this book was profoundly inspired by

Heidegger's philosophy of technology, according to which technology is to be embraced as an abstract mindset, rather than the material aspect of tools and machinery. Citing one of Heidegger's most memorable quotes on this matter: "the essence of technology is nothing technological" (Heidegger, 1982, 35). In the previous chapter, Heidegger's pioneering philosophy of technology was, however, acknowledged as having several shortcomings concerning its capability for framing digital culture and virtual technologies. The least suitable traits of his perspectives in relation to the pursuit of a projectual understanding of virtual worlds are largely encountered in his early work. Among the most problematic of these are:

- its almost completely techno-pessimistic stance toward technology and technological development, and
- the very static and limiting (monolithic) approach to the projectual-possibilities of our technically-mediated being-in-the-world.

When drawing upon this aspect of Heidegger's philosophy of technology in combination with Hughes's theory of technological momentum, what was originally presented as a dark and monolithic understanding of the relationship between technology and society can be somewhat softened. Combining the two, technology can be embraced as the sum of the complex and biunivocal relationship between a functional, objectifying mindset and other human needs that are less reducible and more dependent on cultural factors such as language, cultural heritage, religious views, political orientations, and level of established technological development.

3.1 What is a simulation?

Starting with the etymology of simulation, *simulare* is a Latin verb originally denoting the act of making one thing be similar to another. From this fundamental meaning derive connotations such as to pretend, to falsify, to feign, and to make believe. As Heidegger noted in two of his 1957 lectures (published posthumously under the title *Identity and Difference*), the capability to recognize when things are the same and when they are different from one another is a fundamental precondition for the construction of any ontology.

But what are simulations in the media studies/media philosophy discourse, and what is the most efficient way to understand the relationship between our experience of simulated worlds and the way human beings perceive, understand, and relate to the actual world? Clarifying these questions will contribute to giving context and foundation to the understanding of how ontological overcoming can take place when human kinds of ontologies are "projected" (and are "projecting") into interactive digital worlds.

In a very early attempt to frame the ontological status of simulations, Aarseth associated, in his 1994 *Hyper/Text/Theory*, the concept of simulation with that of cybertextuality. In his theoretical work, a cybertext is specifically understood as a self-changing text in which the organization of the text itself and the possibilities to traverse it are controlled by an immanent cybernetic agent (in Wardrip-Fruin & Monfort, 2003, 777). Aarseth structured his theoretical standpoint on a basic ontological distinction that served as the foundation for the development of his argument. He set up a distinction between experiences that can be understood as being generally fictive and those that take place within a cybertext. According to Aarseth, cybertextuality

> has an element that is not found in fiction and that necessitates an ontological category of its own, which might as well be called simulation.... Simulations are somewhere in between reality and fiction: they are not obliged to represent reality, but they have an empirical logic of their own, and therefore should not be called fictions. (Wardrip-Fruin & Montfort, 2003, 777)

The inadequacy of traditional ontological hierarchies for suitably capturing the qualities of simulated worlds, and their relationships with the world we share as biological organisms, was also recorded by Danish ludologist Jesper Juul. In his 2005 book, *Half-Real: Video Games between Real Rules and Fictional Worlds*, Juul discussed the fundamental ontological ambiguities inherent in games and videogames as follows:

> [Videogames] are two rather different things at the same time: video games are real in that they are made of real rules that players actually interact with; that winning or losing a game is a real event. However, when winning a game by slaying a dragon, the dragon is not a real dragon, but a fictional one. To play a video game is therefore to interact with real rules while imagining a fictional world and a video game is a set of rules as well a fictional world. (Juul, 2005, 1)

In 2003, Frasca proposed the idea that the specific field of game studies had already outgrown the (necessarily) pioneering formalism offered by ludology. However, according to Frasca, structuralist insights are not to be completely discarded; despite all the obvious methodological limitations of a formalistic approach, representational media can in fact be usefully analyzed from that angle. He further specified, however, that the digital platform is not only a representational medium, but is also, if not mainly, a simulational one. This observation was presented in combination with the belief that a rigidly descriptive ludological perspective could not grasp the essential characteristics of changing, negotiable worlds like those that can be accessed through videogames. As a consequence, Frasca considered the development of a performance-based approach a necessary step toward a more accurate and encompassing understanding of games, videogames, or any kind of simulation (Frasca, 2003, 1). Toward that objective, he focused his attention on behavior as a key-term. According to Frasca, "to simulate is to model a (source) system through a different system which maintains to somebody some of the behaviors of the original system" (Frasca, 2003, 2). If we are ready to accept, as Aarseth did, that simulations cannot simply be understood through their output, then we can start to see how a more thorough insight could be achieved when simulations are aptly embraced and explored as mediators that grant interactive access to behavior-based worlds.

Many of the current definitions of simulation embraced by the academic fields of media studies, game studies, and media philosophy, elaborate on the pioneering understanding provided by Frasca. In particular, many of them emphasize the necessary connection between a simulation and reality. This is the case, for example, for Joris Dormans's 2011 notion of iconic simulations (Dormans, 2011), and for the definition provided by Katie Salen and Eric Zimmerman in their 2004 book, where they argue that "a *simulation* is a procedural representation of aspects of 'reality'" (Salen & Zimmerman, 2004, 423).

From adopting the concept of a world as a context[5] characterized by the persistently intelligible qualities of the beings that participate in it, as well as their interrelationships, it follows that, to function as a simulation, a behavior-based world needs to be intelligible. This is necessarily true, at the very least for those for whom the simulation is designed and intended. If a digital experience were to take place in a logical-aesthetic context that was not stably perceivable or that was indecipherable in the way it behaved or responded to user action, it could be argued that the simulation could not meaningfully disclose a (virtual) world to its users. The persistence of its phenomenology and the intelligibility of the causal, spatial, and temporal relationships among beings in simulated worlds are precisely the qualities that allow them to be engaged *as* worlds according to a phenomenological understanding of the term.

Hence, simulations can generally be described as intelligible and persistent, designed interactive ways to disclose complex source systems through less complex, technically mediated ones. The experience of digital simulations, in particular, is structured around (semi)autonomous behaviors. It is crucial to point out that the causal logics and physical behaviors of such behavior do not necessarily have any logical or behavioral dependence on anything outside of the simulation itself. This is not to say, however, that other kinds of relationships do not necessarily exist between the two – chiefly the functional, parental relationship between an actual world and simulated ones.

Another dependency between a simulation and its source system (or systems) that is particularly relevant for the development of my argument is more strictly ontological and will be explored in finer detail in the concluding chapters of this book. To introduce it briefly here, this ontological dependency entails that simulated worlds are set up through processes of analogy with already established ontologies. In that sense, simulations are understood as behavior-based subsets of a wider cultural practice in "an inevitable process of human thought and reasoning" that is called metaphorism. (Kövecses, 2010, x; Bogost, 2012, 74).

It is important to observe here that the behaviors, metaphors, ideologies, and ontologies implemented and integrated in simulations also depend on the simulative possibilities of the medium through which they are accessed and disclosed. Taking place within the current architecture of the digital platform, the virtual worlds of simulations and videogames necessarily inherit basic ontological traits and the expressive possibilities and limitations of computers.

In contrast to the currently popular definitions of simulation as sketched above, in my approach I opt for a definition of simulation that does not rely or focus on its relationships with an alleged reality. This choice has two principal motivations:

- The first motivation stems from the observation that the source system of a simulation can be a simulation itself. The possibility for utilizing a simulation as a source system for another simulation implies that the resulting systems would be progressively more abstract and less recognizable as a descriptive model of behaviors that can be observed in the actual world.
- The second set of arguments that counter the understanding of simulations as detailed models of reality were raised by Juul, who clarified

> that artificial systems classified as simulations can differ strongly from their original source or sources depending on their degree of fidelity, and are usually stylized, meaning that they tend to focus on some aspects of the source system and exclude others, and that they inevitably simplify (Juul, 2005, 170).

The material and structural dependency of simulations on the medium through which they are simulated, and the inescapably subjective understanding of both the source system and how the simulation itself is designed, suggest an analogy between some of the academic frameworks adopted to understand simulations and various perspectives in cognitive sciences and the philosophy of mind. In the words of Neil Stillings, "[c]ognitive scientists view the human mind as a complex system that receives, stores, retrieves, transforms and transmits information" (Stillings et al., 1995, 1). The proposition of a specifically reserved ontological position of simulations, as well as minds, is often advocated as a remedy for what is ultimately the same impasse in the fields of the philosophy of mind and media philosophy. Following this association all the way down the rabbit hole, the most fundamental ontological question that scholars have to face in relation to minds and simulations must be "to what extent is the thing that I am studying a real thing?"

3.2 Simulation and reality

Juul's book, *Half-Real: Video Games between Real Rules and Fictional Worlds*, is specifically concerned with determining the ontological position of videogame worlds, as well as virtual beings and virtual events. As anticipated by the title of his work, Juul argued that the contents of videogames can be understood as both real (if explored from a structuralist, ludologic perspective) and fictive (when embraced from the point of view of their aesthetic and narrative dimensions). This distinction in Juul's work is yet another symptom of the already observed ontological ambiguity that appears to be endemic to simulations when analyzing their elements and behaviors through the lenses of traditional (pre-digital) ontological categories.

According to Juul, regardless of how we decide to understand or define a videogame, some part of it is always more real than another. Interestingly, however, Juul never discussed or included in his argument the material dimensions of what a videogame is (including its hardware, its control mechanisms, or the aesthetic stimuli it is poised to trigger), nor did he explicitly define what he meant when utilizing the term real. In *Half-Real,* the notion of reality is implicitly introduced when discussing the systemic nature of games. On at least a couple of occasions, Juul hinted at the idea that some thing or event being real has to be related to its ontological stability. In other words, the interactive contents of a videogame (e.g., the interactive behavior of a specific game element) can be considered real when they can be considered real when they can be experienced intelligibly and persistently.

Despite openly discussing ontology and the concept of reality in their analytical works on simulation, Juul, Aarseth, and Frasca did not explicitly articulate what it means for anything to be real in their theoretical frameworks. It is my belief that this is indeed a structural deficiency that is not only detrimental to the full intelligibility of the work of these three pioneers of simulation theory and game studies, but also inevitably made the question of the ontological positioning of simulations in relation to the actual world opaque. To ensure that a consistent part of my argument is not undermined by the same ambiguity, I believe it is necessary to provide a workable and sufficiently solid definition of such a fundamental concept on which to anchor all of the assumptions and ideas that follow.

One might be tempted to describe reality as something that is physically present, but this perspective would raise two separate clusters of issues. In the first place, such a definition would reify the duality between the material and the immaterial that held the philosophy of mind captive for nearly four centuries (think of Descartes's concept of the evil genius, discussed in the previous chapter). Second, when trying to identify something real simply as being physically present, the problem of a solid and practical definition is effectively swept under the rug. In defining reality as the property of being present, the nature of the problem we are posing assumes an aesthetic nature, rather than an ontological one. Instead of providing an efficient and agreeable answer to the question, this strategy would defer the definition of what being present means.

The largely overlooked and poorly understood question of "what does it mean for something to be?", and the non-critical acceptance of its associations with the concept of presence, were the springboards for one of the most original philosophical enterprises of the last century, that of Heidegger's question of being. In *Being and Time,* Heidegger raised two critical philosophical arguments against traditional forms of understanding reality either as something that is ontologically stable or as something that has the quality of being present to our senses:

1. *The case against ontological stability:* The first aspect of Heidegger's criticism problematizes the fact that no form of ontological stability can be considered a sufficient criterion by itself to define reality comprehensively. Ontological stability restricts the horizons of the definition of reality to only one aspect of reality (namely its persistence – its conjoined aesthetic and causal endurance in time). This quality of human experience can only offer a standpoint that must be seen as partial and derivative (Heidegger, 1962, 254 / SZ, 211). Heidegger considered this restriction unacceptable and that it did not match the aspiration of providing an absolute and encompassing definition of what it means for something to be real.
2. *The case against presence:* In the second aspect of his criticism, Heidegger paid tribute to Kant and the idea that the human sensory and intellectual equipment is an inherently incomplete tool to relate to reality and make sense of it. The inherent limitation of the human capability to reason, to perceive sensorily, to understand time and space, and even to imagine simply, cannot conjure an extensive definition of reality. Heidegger stated, in these respects, that there is no guarantee that the reality recognized by humans as ontologically present corresponds to the total spectrum of reality. (Heidegger, 1962, 254 / SZ, 211)

Following Heidegger's insights and original phenomenological approach, I propose an understanding of reality as a term that indicates the most basic level of existence, the fundamental background for the perception of phenomena and the development of ontologies. As explained in the previous chapter, for Heidegger a world is always a world *for someone* (something capable of perceiving it, relating to it, and acting within it). The divide between worlds as (post)phenomenological constructs and a reality (a primary and non-sensorily-attainable level of existence) is evident in the distinction that Heidegger posited between the ontological level of beings, presupposing a world experienced and understood via a characteristically human mode of existence, and the ontic level of beings, which is, instead, observer-independent.

The distinction between the terms real and world embraced by this book is a derivation of the understanding of reality outlined above, where the term world is utilized to indicate the way reality is disclosed to the sensory, cognitive, and operational equipment of a certain being. In this sense, when a world is indexed as actual, that world must be recognized for its quality of not being merely potential or possible, but as a world that is subject to property ascriptions and thus can be categorized in ontological structures. In other words, with the premise that sensory mediation must, nevertheless, be considered as mediation, reality could be succinctly defined as that which is immediate.[6]

Having proposed a workable definition of what the terms reality and world mean in the context of this book, the last preparatory step before pursuing a rigorous exploration of the effects of virtual worlds on human kinds of ontologies is an explanation of what exactly is meant by the adjective virtual.

3.3 What is virtual?

Etymologically, the adjective virtual derives from the Latin *virtualis*. *Virtualis* is not a classical Latin term, but it is a late-medieval neologism whose existence became necessary when Aristotle's concept of *δύναμις* (*dynamis*: potentiality, power, quadrate) had to be translated into Latin (Van Binsbergen, 1997, 9). The concept of potentiality at the etymological foundation of the adjective virtual provides the background for understanding why, at least in one of its interpretations, it is used to indicate the latency of certain possibilities inherent in a specific artifact, combination of artifacts, or state of things.

By definition, every simulation is characterized by the potentiality of virtual alternatives to its current state. An evident example of this latency can be found in the editorial arrangement and the consequent structuring of the literary content in Raymond Queneau's 1961 literary simulation, *Cent Mille Milliards de Poèmes* (see Figure 3.1). All of the unexpressed potential of the combinatorial content in Queneau's book is virtual, while the only non-virtual combination (the one corresponding to its current arrangement) is commonly indicated as actual. As a direct consequence of this observation, a simulation that is not active or does not have a definite, actual configuration (e.g., a closed copy of *Cent Mille Milliards de Poèmes* or a board game still packed in its original box) has the range of virtual combinatorial possibilities completely available.

This first understanding of virtual is particularly apparent when experiencing digital media content, content that is characterized, paraphrasing the perspectives of Lev Manovich, by its modular, interactive, and self-organizing qualities. The combinatorial nature of digital content is evident, for example, in the functioning of software such as videogame engines and level editors. This kind of middleware offers videogame designers, level-designers, and digital artists the tools to configure finite sets of game elements from existing assets or databases into game worlds or sub-worlds (levels).

When users (players) experience, traverse, and manipulate computer-simulated worlds, instant after instant, their screens display the present (actual) state of that particular world or sub-world. Each state of a digital world has the inherent possibility of developing and changing into innumerable other potential configurations that have a perceivable logical (causal) connection with the present one. All the hidden paths and all the unexpressed possibilities offered by virtual worlds exist virtually within the way the software was designed and the possibilities offered by its affordances.

This first kind of digital virtuality can take place in the form of either prescriptive branches or more flexible and interactive gameplay options. With regard to the latter possibility, as an example, it could be useful to think of the multiple different ways to reach the end of a level in platformer videogames, where several alternative paths are viable for the player and multiple different gameplay approaches allow for progress in the game. Hybrid combinations of the two interaction-design approaches mentioned above are also possible. These strategies are particularly conspicuous and frequent in the interaction with videogames that are focused on delivering a narrative experience – for example, in *Sega*'s *Shenmue*, *Capcom*'s *Resident Evil 4*, *Bioware*'s *Mass Effect* series and *Quantic Dreams' Heavy Rain* – where sections of gameplay that are more dominantly action-oriented are punctuated by sections that are designed specifically to deliver narrative content. In parts of a videogame that are more focused on narration, the player is asked to choose between different branches of the plot, usually in the form of ergodic dialogue options or quick-time events.[7]

Remaining with this interpretation of virtuality, the virtual quality of videogame content is particularly easy to recognize in *Gua-Le-Ni; or, the Horrendous Parade*, an Apple iPad and iPhone videogame that I designed and developed (henceforth referred to as *Gua-Le-Ni*) (Double Jungle S.a.s., 2011–2013). The world of *Gua-Le-Ni* takes place somewhere in Great Britain and is set temporally in a fictional reinterpretation of the Age of Discovery. In *Gua-Le-Ni*, the player is given the role of an aspiring scholar who is instructed by an old, befuddled British zoologist on the finer points of combinatorial taxonomy. On a dark wooden desk lies an extraordinary book, a leather-bound bestiary populated by bizarre, finely drawn paper creatures that allegedly inhabit the new world. Similar to the combinatorial monsters of head-body-tail books that we might have played with in our childhood and to the creatures recounted in legends and myths, the paper beasts of *Gua-Le-Ni* are chimeras, impossible assemblages of real animal parts (Gualeni, 2015).

The combinatorial paper creatures of *Gua-Le-Ni* walk hurriedly across the illustrations of the bestiary from their right margins to their left ones. From the point of view of the player, the main goal of the game is to recognize the components of the fantastic creatures and their relative order before the creature manages to completely traverse an illustration and escape from the book (which constitutes the game over condition). Encouraged by the unwieldy mentor, the player pursues this purpose by quickly rotating, moving, and spinning toy cubes with pictures of animal parts printed on each face of the cubes. A paper beast is correctly identified – and thus prevented from escaping the old book in which it belongs – when the player manages to match the illustrations on the top faces of the taxonomic cubes with the paper beast currently in play (Gualeni, 2015).

During a game of *Gua-Le-Ni*, only one combination is displayed at any given time. The individual beast that is presented to the player is generated by an algorithm that I designed, chosen from thousands of possible beasts to fit certain requirements of difficulty and solvability. In other words, each specimen walking across the game is a combinatorial being, an instance of a virtual field of monstrous possibilities. In Figure 3.2, the CA-BIT-DOR-STER (a four-module creature with the head of a camel, one body part of a rabbit, followed by a mid-section of condor, and concluded by a lobster's tail) is the "actual" beast, while tens of thousands of analogous creatures are to be considered virtual in the sense that, in the instant that the screenshot was taken, they were possibilities within the combinatorial system that remained unexpressed.

A second interpretation of the adjective virtual was presented by Pierre Lévy as not in opposition to actual in the sense of current (presently existing) but to actual in the specific meaning of "pertinent to the world humans are native to" (Lévy, 1998, 14).[8] This ulterior aspect of the definition of what virtual means cannot be understood in the restricted context of a single, self-enclosed world (as was the case with the first meaning), but requires the concurrent existence of more than one world, at least one of which needs to be indexed as the actual one. In this second meaning, Queneau's book is not virtual, but is objectively actual. Proposing a similar interpretation to the one just outlined, Michael Heim defined a thing as being virtual when that thing is actual "not in fact, but in effect" (Heim, 1994, 109–110). Interestingly, the

Figure 3.2 A screenshot taken from *Gua-Le-Ni; or, the Horrendous Parade* (Double Jungle S.a.s., 2011–2013), showing a CA-BIT-DOR-STER parading across the screen (iPad screenshot used with authorization)

beasts that are not currently in play during a game of *Gua-Le-Ni* are virtual in both senses.

While playing *Gua-Le-Ni*, the digital tabletop and the simulated events that take place on it (the pages turning, cubes realigning, paper beasts being blown away by sudden gusts of wind, etc.) reorganize themselves on the basis of player interaction. The persistently intelligible elements of the game provide aesthetic feedback in ways that are consistent both in their internal logic and in their responses to player interaction. Proceeding from these observations and based on the definition of a world as given in Chapter 1, *Gua-Le-Ni* (along with any videogame or simulation sharing analogous characteristics) must be seen as disclosing a virtual world to be experienced and manipulated by players. In general, and in line with Manovich's observations, virtual worlds are worlds that are characterized by the qualities – largely dependent on their digital mediation – of being modular, interactive, and self-changing. Digitally mediated simulations grant access to the perception of and interaction with such worlds that, regardless of their virtual constitution, disclose interactive experiences that are perceptually and cognitively effective.

The digital entertainment industry attempted, from its onset, to disclose worlds that are as worldly as possible, in terms of aesthetic consistency, granularity, and cognitive unobtrusiveness. This game design attitude, aimed at maximizing the appeal and the intuitive accessibility of virtual, playful experiences, is commonly indicated in the field of game studies as *mimesis*, in analogy with the attribute of tragic theatre that Aristotle recognized as eliciting empathy and necessary in the pursuit of catharsis and truth (*μίμησις*). In the first decade of the twenty-first century, the videogames industry creatively converged on ludic metaphors that relied on simplified versions of the traditional physical understanding of the world and on behavioral and representational conventions of Western societies. It is not a coincidence that *Super Mario Bros.* is "read" from left to right, that the player is explicitly asked to pursue values ascribable to specific worldly ideologies,[9] and that the physical qualities of the world that *Super Mario* inhabits are intuitively intelligible to human players in analogy to the ones the players themselves experience in their everyday relationship with the actual world (a couple of notable examples of this cognitive association can be observed in the way gravity works in the Mushroom Kingdom, or the fact that fire and lava are implicitly introduced in the videogame as hazards for the anthropomorphic playing character).

At the same time, it must be noted that the worlds that can be experienced in videogames tend to have qualities and behaviors that are largely unworldly. Unlike what happens in *Gua-Le-Ni*, to use the game as an example again, in the world that we share as biological organisms paper creatures do not magically come alive and start walking through illustrated landscapes, paper apples do not grow from paper trees, and time cannot be halted by turning the pages of a book. In *Super Mario Bros.*, and differently from the real world, creatures can have additional lives, double jumps, and can shoot fireballs.

Since the very first vectors drew rudimentary spaceships on a screen, videogames have presented experiential systems that are often significantly different from those that humans have commonly established in their relationships with the world throughout the history of culture. After more than 30 years of technological development, recent videogame productions still openly (and in a willingly escapist way) defy traditional worldviews, making phenomenological unworldliness a central theme of their creative agenda. In this sense, videogames can be seen as embracing strategies of aesthetic estrangement in ways that are similar to modernist movements such as Dadaism, Surrealism, Situationism, or Russian constructivism. Titles like *Super Paper Mario* (Intelligent Systems, 2006), *Portal* (Valve Corporation, 2007), *Crush* (Zoë Mode, 2007), *Echochrome* (Japan Studio, 2008), and *Portal 2* (Valve Software, 2011), just to mention a few, specifically violate the traditional understanding of space, actively granting their players the chance to reinterpret its dimensions, continuity, and homogeneity.

The understanding of time as a homogenous, continuous series of causally linked nows is another idea that was often identified in the history of philosophy as the foundation for the structuring of traditional worldviews. This classical understanding of the dimension of time, which is also a component of classical mechanics, is flagrantly violated in the virtual experience of time offered by games such as *Legend of Zelda: Majora's Mask* (Nintendo EAD, 2000), *Blinx: The Time Sweeper* (Artoon, 2002), *Prince of Persia: Sands of Time* (UbiSoft Montreal, 2003), and *Braid* (Number None, Inc., 2008). These are just a few instances of a trend in game design that empowers the player to actively create, manipulate, and relativize established ontological assumptions. Challenging concepts like duration, reversibility, and causality are at the very core of the innovative game-mechanics of several titles in modern interactive digital entertainment.

Other, and conceptually similar, examples of digital insubordination could be listed in relation to a multitude of ontological topics and categories other than space and time – for example, interactively subverting concepts like personal identity, infinity, or the problem of the universals. In modern videogame development, the laws of physics and the stability of most categories of pre-digital human kinds of ontologies are not only challenged, but are themselves increasingly interactive and modifiable elements of play. This book does not consider such infringements as a defining trait of videogames; rather, it considers the interaction with digitally mediated simulations a particularly evident, popular, and immersive way in which the experience of virtual worlds fosters the distortion, extension, and fragmentation of human thought and experience.

4
Thinking with Virtual Worlds

In his philosophy-inspired work, Argentine writer Jorge Luis Borges often suggested imaginative alternatives to the customary Western understanding and representation of the world. In the short story, "The Analytical Language of John Wilkins," for example, Borges informed his readers that, in "a certain Chinese encyclopedia" (the *Celestial Emporium of Benevolent Knowledge*), the animals are divided into:

> (a) belonging to the Emperor, (b) embalmed, (c) tame, (d) suckling pigs, (e) sirens, (f) fabulous, (g) stray dogs, (h) included in the present classification, (i) frenzied, (j) innumerable, (k) drawn with a very fine camelhair brush, (l) et cetera, (m) having just broken the water pitcher, (n) that from a long way off look like flies.[1] (Borges, 2001, 231)

A deep ontological chasm separates the fictional taxonomy of the Chinese encyclopedia and the scientific cataloguing of fauna initiated by Carl von Linné halfway through the eighteenth century. In the face of the incongruence between the two epistemological approaches, Michel Foucault, French philosopher and social scientist, commented that Borges' fabulous categorization demonstrates, through the "exotic charm of another system of thought," the "limitation of our own, the stark impossibility of thinking *that*." (Foucault, 1994, xv)

Carol J White has noted that, within Borges' taxonomy, "[a]n animal could change categories moment to moment or fall into more than one of these 'species' at once depending on its relation to the viewer, current activity and so forth" (White, 1996, 148). Drawing upon Heidegger's understanding of the "temporality of Being," White emphasized that the ontology underlying the categorization presented by the Chinese encyclopedia diverges radically from the aspiration to the universality, consistency, and stability of knowledge that characterizes Western metaphysics and underlies scientific inquiry.

White considered it to be particularly illuminating to observe such a discrepancy through the lens of Heidegger's thought, focusing her attention on the defiance of the temporal stability that characterizes traditional ontologies. Presenting the reader with taxonomic categories that can be impermanent or accidental ("belonging to the emperor," "frenzied," "having just broken the water pitcher"), the fictional ontology that can be inferred in the Chinese encyclopedia derided the aspirations of Western metaphysics for describing reality in a manner that is extensively consistent and temporally stable. The Chinese encyclopedia categorizes animals in a way "that it involves no enduring Being of what-is united across past, present, and future or specifiable in every 'now'" (White, 1996, 148).

The divergences between the *Celestial Emporium of Benevolent Knowledge* and Western thought are not limited to temporality. As an especially revelatory example, Borges' fantastic taxonomy distinguished certain beasts from others based on their looking like flies "from a long way off," offering a classification that, instead of pursuing absoluteness, is relative to the spatial rapport between the animals and the observer. If this were not enough already, in the previously mentioned passage from "The Analytical Language of John Wilkins," Borges stated that a possible metaphysical categorization of animals could rely on their being "drawn with a very fine camelhair brush," violating the traditional ontological hierarchy between things and their representations.

In synthesis, Borges' insubordination to traditional ontological conventions consists in providing fictional, imaginative alternatives to the way humans customarily give order to the world and functionally relate to it. The exhilarating and thought-provoking ontological possibility evoked by the *Celestial Emporium of Benevolent Knowledge* has additional qualities that can be seen as inherited by the media through which they are articulated. In the case of Borges' fictional worldviews, these qualities are shaped around the limitations of text, including (most evidently) their being: bound to certain semiotic codes; unchanging in time; non-interactive; and impossible to experience phenomenally.

This chapter focuses precisely on framing the expressive constraints of different media forms and their possibilities to be employed in the overcoming of traditional human kinds of ontologies. Before tackling this central philosophical task, I find it necessary to posit a distinction between the concepts of simulation and fiction that is foundational for constructing an ontological perspective on digital media.

From the postphenomenological standpoint embraced in this book and following the observations in the previous chapter, digital simulations are recognized as artifacts capable of granting access to intelligible, perceptually stable, self-changing, and interactive virtual worlds. In a way that could be associated with simulations, traditional media of communication can disclose experiences that are also nominally independent from the actual world (for example, in the case of a science fiction novel, or an abstract painting), but that are experienced on the basis of subjective imagination and interpretation. As such, I argue, traditional forms of expression cannot provide experiences that emerge from establishing relationships with an objective world. The disclosure of such experiences is, currently, the sole domain of simulations, setting simulations apart from the fictional alternatives to the world presented by traditional media. I believe that simulated experiences hold a different ontological status from those offered by traditional media, whose influence cannot but remain confined within the boundaries of fiction.

4.1 Pre-digital media forms and their ontological influence

Plato, whose thought and work lie at the very core of Western metaphysical tradition, understood art as an imperfect copy of something that humans encounter in the world (Plato, *Republic*, 605). The world, in Plato's Theory of Forms, is itself a flawed material instance of the perfect and eternally unchanging world of ideas. Consequently, Plato viewed art as a doubly flawed imitation (*mimesis*) of certain forms or ideas, which art could never logically aspire to capture or understand. Artists were, consequently, to be considered lowly craftsmen, and the more their specific craft was involved with the manipulation of materials, the lower their social status should be (with musicians being the most socially valuable artists and, presumably, architects and sculptors being the least esteemed).

Surprisingly, it was also within the Platonic tradition that alternative perspectives to the Platonic, mimetic conception of art were first proposed. In Plotinus' *Enneads*, for example, works of art were not considered to poorly mimic aspects of the already imperfect material world, but were understood, rather, as operating in parallel to the world by executing and expressing the artist's original vision and presenting new worlds (Plotinus, *Enneads* I 6.3; V 8(31).1). As a consequence of this shift of perspective, artists were no longer perceived as earthly artisans, but rather as individuals whose creative and socially valuable activities could be compared to those of the philosophers.

The change in the understanding of the social role of art from futile *mimesis* (imitation) to worthy *poiesis* (creation) is particularly obvious in the tradition beginning with Romanticism, where the figure of the artist replaced God (both metaphorically and figuratively) as the creator and originator of new worlds. Gregory L Ulmer argued that modernism, especially in the use of new techniques such as collage and montage

> does not reproduce the real, but constructs an object...or rather mounts a process...in order to intervene in the world, not to reflect but to change reality. (Ulmer, 1983, 86 in De Mul 2010, 155)

The artistic strategies adopted by Dadaism, for example, defied the figurative and representational canons that preceded modernism. The works produced by artists associated with Dada often consisted of juxtapositions of incongruent elements that did not constitute a recognizable or familiar scene. The incongruence between their works and our everyday, proximal experience of the actual world was not only relative to the scale of their elements, their type, or their belonging to heterogeneous contexts, but also encompassed a discordant variety of representational techniques.

Dada's artistic objective was to weaken what the movement saw as the totalizing grip of means/end rationality on Western culture. This objective was pursued by subverting traditional aesthetics, challenging traditional representational canons, and evoking aesthetic possibilities that were not compatible with the everyday experience of the world (Gualeni, 2014a). The general insubordination of modernist movements to the univocality and stability of the Western tradition of thought was explicitly avowed in their manifestos and by the deliberate use of artistic expression as a catalyst for social change.[2]

Dadaism's rebelliousness against social and artistic conventions was a precursor to, and an inspiration for, radical avant-garde currents, such as Surrealism (in general, but especially in their playful and interactive installations), German Expressionism (particularly in Expressionist cinema), and Situationism. Theories on intellectual liberation via aesthetic appreciation arguably played a vital role in the social agendas of several strands of the artistic avant-gardes of the twentieth century. Such theories left conspicuous traces in both the philosophical and the literary production of the same period.

Philosophers and aesthetics scholars of the last century, such as Hans Robert Jauss, Jacques Ellul, or the later Heidegger, explicitly structured their work on technology as a reaction to the commoditization and alienation of human existence that they observed in the mechanization of work and transportation, and in the general diffusion of technological mediation in social processes and practices. As a possible alternative to the technological mindset, these scholars often proposed the free encounter with art as a means to achieve liberation from both the imposition and the limitations of Western, scientistic thought. Albeit passively experienced, traditional forms of artistic expression were considered capable of detaching people from their everyday and functional existence and leading them into a wider, freer realm of sensory appreciation.

This particular interpretation of the social role of art is by no means a product of the twentieth century. It is, in fact, a recurrent trope of the philosophy of Western art that the attentive reader may have already noticed surfacing in this book on at least two occasions: when mentioning Aristotle's understanding of *mimesis* as a way to pursue catharsis in tragic theatre (in Chapter 3); and when touching upon Plotinus' perspectives earlier in this chapter. In the twentieth century, however, two parallel ideologies contributed to the spreading and dominance of the interpretation of art (and the appreciation of art) as liberating activities:

- The first ideology focused on the removal of the individual from his or her customary and functional social context. This interpretation of the sociocultural role of art relied on the alleged possibility of the work of art to stimulate a process that Russian critic Viktor Shklovskij defined as one of "estrangement" or "distancing." According to this ideological stance, social and artistic production was considered capable of revealing familiar, everyday details from unusual, foreign angles. The work of art could, thus, reveal the world to the "estranged" observer as something alien, something that is encountered for the first time, rather than something that is simply recognized in its habitual everydayness. Such a purpose was pursued in a particularly explicit fashion by the Russian constructivists through bizarre photographic experiments, as well as installations, paintings, and prints that involved various (and often dynamic) combinations of mechanical parts, geometric cardboard shapes, and unfamiliar typographical arrangements of Cyrillic letters. The deconstruction of the familiar forms and the customary arrangements of things were meant to impose, according to Shklovskij, a new perspective on things: a perspective suggestive of "the possible" that relies on the belief that cultural production is capable of revealing the possibility that the world can be different from what it currently is. This position was shared by Heidegger's early stance in relation to the cultural role of the work of art (Gualeni, 2014a).

- The second ideological stance took an approach to defining the sociocultural role of art that was largely antithetical to the first. The salvific and emancipating potential that this perspective ascribed to art related to its original meaning of doing discussed at the beginning of the first chapter. As such, this second perspective was grounded in the belief that the practical crafting of an (art) object or a product could lead the individuals involved in the creative process to the establishment of a more direct and genuine engagement with themselves and with the world. Similar to the first ideology, however, this second perspective also opposed the encroaching mechanization of the world along with the consequent establishing of adulterated and insincere relationships between people and distancing people from what is understood as genuine engagement with the world. This ideological stance often proposed manufacturing (the direct involvement of the individual with cultural and artistic production) as a remedy to the already mentioned means/end rationality. Art was considered to be infused with the power not only for realizing the liberating, creative potential of human beings as creators, but also for allowing things to manifest themselves in their real essence for the potential recipients of the art. According to an analogous perspective, the emancipative capabilities of artistic expression were also believed to reside in the bodily (non-mediated) and active re-appropriation of the world, and in the consequent recapturing of what was believed to be authentic humanity.[3]

Twentieth century theories of aesthetic liberation from the habitual relationship with the world (or rather with its functional objectification) inevitably attributed to traditional media of representation the capability of having deeply liberating effects on human thought and human behavior. The literary and theatrical productions that aligned with the Pataphysical tradition, together with Hans Bellmer's Surrealist photographs and all instances of absurdist fiction, are particularly evident cases of how such awareness distinctively affected cultural production in the last century.

The capability of presenting fictional alternatives to customary and widely recognizable ways that human beings relate to the world in their everyday existence does not solely pertain to literature, theatre, painting, or to Dadaist techniques, such as those of collage and photomontage. Besides the case of German Expressionism, David Lynch's films constitute an often cited and manifest logical-aesthetic defiance of traditional ontological and narrative conventions. Similarly, the films of Katsuhito Ishii are obvious examples of how moving pictures are not an exception among traditional forms of media in terms of their capability to evoke fictional worlds that are incompatible with the one we share as biological creatures. The same logic applies to ballet. In the last century, ballet proved capable of choreographing fictional worldviews that were often drastically incoherent compared to those experienced by human beings in their customary and unabridged involvement with the actual world. One famed example of this were the abstract, moving geometries of Oskar Schlemmer's 1922 *Triadic Ballet*.[4] These observations of cinema and dance are equally valid for all other traditional forms of media, which consequently need to be recognized as capable of fictionally representing worldviews, as well as their alternatives, distortions, and subversions.

As was briefly discussed at the beginning of this chapter, the central theme of Borges' writing pivoted on the capability of literature to suggest alternative ways of understanding, categorizing, and imagining worlds. As can be surmised by the title of his most critically acclaimed book, *Fictions*, Borges' work is composed, for the most part, of literary infringements of traditional ontological conventions. His novels can be described, using Borges' own words, as "[t]he conjunction of a mirror and an encyclopedia," a deliberately deceptive use of text in the mediation of (ontological) thought (1994, 17). Borges' novels and short stories demonstrate the possibility for fictional writing not only to re-present the world, but also to replicate, fragment, and distort its representations. In other words, text, and any traditional form of mediation, not only describes[5] what is actual, but can fictionally – and thus subjectively – evoke alternative worlds and worldviews alike.

According to the perspectives adopted in this book, the qualities of textual mediation outlined above would disqualify it (and any other form of traditional, representational media) from being capable of influencing and altering human kinds of ontologies. The reason for this disqualification lies in the very definition of ontology: a particular way for a being to understand and organize rationally its experience of one or more worlds. My argument is that the experiences and sensations evoked by forms of mediation that are simply representational do not amount to a world according to the definition adopted by this book. This is because the things and events represented through traditional media are not characterized by the possibility for establishing stable perceptual, cognitive, and (possibly) operational relationships with their recipients. Put somewhat more simply, traditional media cannot provide the effective materialization of worlds and, as such, cannot be expected ever to engage a subject at the ontological level. The subjective and representational horizon of traditional mediation defines the cognitive and epistemological boundaries of the concept of fiction. Heidegger presented an analogous perspective on the limitations of representational media in the "The Memorial Address," where he explained that, through these media, people are given "the illusion of a world that is no world" (Heidegger, 1966, 48).

Having observed that representational expressive forms cannot be considered mediators of human thought and experience, it is important also to mention the fact that, in the past century, less conventional and more flexible uses of the textual medium were attempted and tested. Combinatory literature and gamebooks managed, for example, not only to overcome some of the traditional limitations of the textual medium – such as the linearity and stability of textual compositions – but also pioneered the exploration of the boundaries of the simulational mindset. Such literary experiments are often cited as tangible expressions of a society whose values, expressive needs, and technological environment were changing inexorably.

Frasca has observed that the full power of simulation was unleashed from its technical (textual–editorial–representational) limitations with the invention of the computer (Frasca, 2003, 2). Computers grant access to virtual worlds – worlds that are not real in fact, but are real in their experiential effects. The ontological stability of the virtual worlds disclosed by digital simulations and videogames, as well as the level of detail of the aesthetic stimuli these digital media offer in comparison with traditional media, have often been viewed by theorists and philosophers as decisive factors in a fully fledged cultural shift. Michael Heim, for example, observed that interaction with digital simulations is not bound to subjective evoking by means of fictional representations. Rather, this interaction involves the whole spectrum of processes through which humans relate to worlds (Heim, 1994, xiii). In accordance with this idea, Frasca also maintained that, with the advent of the digital medium, the logical, aesthetic, and interactive possibilities of simulation increased immensely, as did the complexity of the systems to which they grant access. Because of these new technical affordances, according to Frasca, the encounter between the simulational mindset and digital mediation opened a wider horizon of possibilities for mankind. According to the perspective proposed in this book, the new horizon that virtual worlds are opening up is an artificial (and potentially unworldly) experiential field in which human ontologies can project, fragment, distort, and extend.

Borrowing from Herbert Marcuse, traditional media are seen as extending and ensuring the prevailing *Lebenswelt* (the German for lifeworld: the world that subjects may experience together) without altering its existential structure, that is, without envisaging a new mode of "seeing" (originally written in relation to the scientific method; Marcuse, 1991, 165). My argument is that, differently from traditional media (which rely on subjective imagination and largely on the passive assimilation of content), interactive digital media can objectify ontological alternatives to the *status quo*. From this standpoint, the virtual worlds supported and disclosed by computers can be interpreted as (heuristic) ontological instruments. As such, a thorough understanding of virtual worlds as mediators of human thought and experiences, and of their effects on human kinds of ontologies, requires a perspective capable of understanding digital technology philosophically and anthropologically. The next section begins the articulation of a perspective on the philosophy of technology that, borrowing from the postphenomenological tradition, pursues that specific objective.

4.2 From "subjectivity" to "projectivity"

Toward the end of the nineteenth century, the German philosopher Ernst Kapp was among the first to propose that all forms of technology were artificial extensions of the human organism. He put forward a fundamental understanding of technology built around the concept of "organ projections," the belief that technologies are conscious (or even unconscious) actualizations of original functions of human organs (Kapp, 1877). Through technology, according to Kapp, humans can supplement the shortcomings and the malfunctions of their native organism. The outlined, functional aspect does not, however, constrain the effects and roles of technologies to extending our capabilities to perceive, communicate, and operate in the world (or worlds). Rather, our artificial extensions are also recognized as cognitive instruments. Elaborating on Kapp's original intuitions, Verbeek added another dimension to the human capability of "objectifying" themselves (or parts of themselves) in material apparatuses. Through the technological externalization of some of their functions, human beings also reveal a technological domain of self-discovery (Verbeek, 2013, 232).

Starting with the Enlightenment, the scientific process structured (with increasing granularity) a mechanistic understanding of nature. Observed from this perspective, Kapp's reflections appear to be working in the opposite direction to the Enlightenment project. Kapp's work can, in fact, generally be understood as having attempted to understand the mechanic in terms of the organic. Starting from the recognition of the biological origins of technically-fulfilled needs, Kapp did not posit a dialectical separation between the natural world and the artificial world, or between the realms of technology and society. To elucidate what it means to understand digital technology as an integral part of who we are as biological creatures and social beings, it is useful to outline the qualities of technologies as taxonomically organized in the 1950s by German philosophers Hermann Schmidt and Arnold Gehlen.

In his 1954 essay, "Die Entwicklung der Technik als Phase der Wandlung des Menschen," ("Technological Development as a Phase of the Transformation of Man"), Hermann Schmidt elaborated his understanding of the relationships between the organic and the technical. In his text, he identified three progressive stages in the development of any form of technology. The first stage was that of tools. The perspective on technological development embodied in this stage can be seen as the dominant perspective in Kapp's work and was closely reminiscent of Heidegger's understanding of equipment presented in his writings on technology. For Schmidt, tools were simple technologies, intuitive in their use, that could be understood as direct extensions of either the physical (e.g., a hammer or a pair of glasses) or mental (e.g., an abacus or a paper notebook) capabilities of human beings. According to the understanding of perception and cognition proposed by French phenomenologist Maurice Merleau-Ponty, in his 1945 book *Phénoménologie de la perception* (*Phenomenology of perception*) familiarity with tools and efficiency in their use emerged from the incorporation of the affordances of the said tools in the "body schema" of a certain subject. For Merleau-Ponty, the body schema was the pre-cognitive organizational structure that determined how beings understood their bodies and created relationships with the world they inhabited (Merleau-Ponty, 1962; Carman, 1999, 218–223). Concluding this synthetic description, tools were understood in this taxonomy as simple, passive instruments that could not function unless incorporated and actively employed by (human) beings.

The second stage of technological development proposed by Schmidt was that of machines. Machines were still relatively simple instruments, but, when compared with tools, they could be distinguished by a higher degree of autonomy in relation to their functional employment. A machine functioned according to certain planned behaviors but still had to be operated to some extent by a human being. Unlike the directness of tools, machines were semi-autonomous physical representations of their design (Coolen, 1992, 34).

Finally, the third stage in the development of technologies as understood by Schmidt was that of the automata; supremely autonomous technological artifacts that were characterized by a greater autonomy in both their physical operation and the purposeful deployment of their machinery (Verbeek, 2013, 232). The automaton realized the functional objective of technology, rendering the human operator redundant (albeit still necessary to its design and maintenance).

In his 2010 book *Cyberspace Odyssey*, De Mul elaborated on the third stage of Schmidt's taxonomy. Inspired by the writings of Alan Turing, De Mul presented computers as yet another stage in the process of the development of technology. Computers are universal machines; electronic systems that cannot be understood as the autonomous or semi-autonomous externalization of a certain need or project, as was the case with general machines or automata. De Mul explained that the uses and applications of computers are not the same as the functioning and the material configuration of the computer-machine (its hardware). Rather, the uses and applications of computers are more suitably identified with their software (De Mul, 2010, 114). The variety, flexibility, and individual nature of each software application allow us to characterize the digital medium with a definite, univocal, socioeconomic momentum.[6] De Mul proposed a vision according to which computer applications, unlike pre-digital technologies, cannot establish stable and conventional relationships with societies.

American journalist Nicholas Carr discussed this peculiar quality of digital media in very similar terms in his 2010 book, *The Shallows – What the Internet is Doing to Our Brains*. As a significant example, Carr addressed the fast and ethereal nature of businesses based on computer applications, writing that "[a]ll it takes to render a thriving online business obsolete is a sharp programmer with a fresh idea" (Carr, 2010, 157). In contrast to the previously outlined techno-pessimistic critical horizons, this perspective might give way to technological optimism in the sense that it does not envisage a totalizing sway of artificial systems. Rather, it presents an impermanent state of both technological development and its opportunities for influencing society. According to this perspective, digital technology and digital applications could, in principle, allow for more transparency, a more democratic and flexible access to resources and information.

One of the first academics to argue that the specific sociocultural changes brought on by computers need to be understood anthropologically, in terms of quality (rather than quantity) of operations, was philosopher and computer linguistics pioneer Margaret Masterman (Willard McCarty, 2012, 114). Writing in 1962, Masterman encapsulated her vision concerning the potential of the digital medium for expanding and transforming the human world in the metaphor "a telescope for the mind" (Masterman, 1962, 38). Her intuition established an analogy between the advent of computers and the impact of the developments of optical technology in the seventeenth century on the knowledge of cosmology. Masterman further explained that, by extending the perceptual scope and reach of mankind, computing does not simply and neutrally bring formerly unknown and never before experienced things into view, but also forces society into an epistemological crisis from which novel worldviews and new ways of understanding ourselves arise.

Half a century before Masterman, both Heidegger and Karl Jaspers laid the foundation for an understanding of technology that is inextricably linked with the ways that humans structure their relationship with reality. In the last 30 years, these ideas have been valorized, criticized, and expanded upon by Don Ihde and the postphenomenologists in the general context of the philosophy of technology. Adopting a postphenomenological approach to digital mediation, my specific claim regarding the core of the cultural shifts brought about by digital technology is that the qualities of digital technology allow human beings to objectify and overcome some of the phenomenological, operational, and ontological boundaries that characterize pre-digital thinking. The same restrictions are, in my view, manifest in the way traditional forms of mediation frame and shape information, and in the way Western values and worldviews are materialized in our technological lifeworld.

In all the perspectives of the philosophy of technology outlined in this section, humans are understood as characterized by their being "artificial by nature." This has been the case, according Plessner, since the very moment *Homo habilis* manufactured the first stone tools (Plessner, 1975, 385). Not unlike Kapp's philosophy of technology, McLuhan's understanding of technology as the "extensions of man", or Plessner's idea of a "natural artificiality," American theorist Jay Bolter argued that human subjectivity has been dynamically integrated and augmented by artificial means from the very onset of civilization, and is now natively integrated with computers as the dominant form of mediation in Western cultures (Bolter & Grusin, 2000, 41–44).

In a paper that explored the concept of immersion in simulations, cultural historian Erkki Huhtamo also pointed out that technology has always been involved with humanity and has, as such, always effectively been second nature to mankind. It is particularly interesting to notice that, for Huhtamo, the dynamism in the relationship between humans and technology in the digital era is particularly evident in the fact that the design of digital interfaces no longer seems to pursue integration (a symbolic association with similar functions in the actual world), or transparency (the quality of being perceptually as close as possible to a non-mediated experience). However, these interfaces also explicitly rely on functional conventions that are inherited from the tradition established by previous interactive technologies "simply because it is not felt to be in contradiction to the 'authenticity' of the experience" (Huhtamo, 1995, 171).

After these observations, it appears that in the present age of human enhancement, people no longer simply design their lives from an existential standpoint, but do so from a biological and a phenomenological one. This form of projectuality is a process that, by definition, cannot take place without ontological consequences. The mechanistic technologies reflect the purpose of the rational domination of a world that is objectified and reduced to a system of usable resources, whereas digital and virtual technologies focus on the possible – on the many ways the world could be designed, re-designed, and manipulated. Discussing the social impact of digital technology from an analogous perspective, Vilém Flusser wrote that:

> [W]e begin to liberate ourselves from the tyranny of an alleged reality. The slavish attitude, with which we, as a subject, approach objective reality in order to master it, has to give in to a new attitude, in which we intervene in the fields of possibilities in- and outside us, in order to intentionally realize some of these possibilities. From this perspective, the new technology means that we are starting to raise ourselves from a subjectivity into a projectivity. We are facing a second birth of mankind, a second *Homo erectus*. And this *Homo erectus*, who plays with chance in order to intentionally transform it into necessity, may be called *Homo ludens*. (Flusser, 1992, 25)

With the proliferation of digital media and the increasing production of interactive virtual worlds, human kinds of ontologies can, in a reciprocally influential relationship with digitally mediated simulations, extend and fragment into worlds that were previously (pre-digitally) inaccessible. As a consequence of the diffusion and assimilation of digital media in social processes and practices, human ontology shifts into an increasingly technically-mediated context; a labyrinth woven by men to be deciphered by men, with a rigor, as Borges warns, that is not that of angels but rather of chess players (1994, 34).

4.3 Virtual worlds as "technologies of the self"

Generally, the perspectives on technology espoused in this book accept that humans are beings that are artificial by nature. The introduction of any form of technology into social processes and practices is understood as constituting a factor of cultural change because of its inherent capability for altering and extending the horizon of possibilities for humans to perceive worlds and relate to them both intellectually and operationally. My understanding also embraces technological development as a way for human beings to materially objectify their worldviews, needs, and aspirations. I find this perspective of philosophy of technology particularly poignant as, by allowing people to expressing themselves through technical artifacts, technology also discloses specific forms of self-reflection and self-discovery. This auto-gnostic aspect of how human beings extend and objectify themselves, their ideas, and their desires in technical artifacts and systems is present in the work of several academics in the field of the philosophy of technology. Donna Haraway tersely stated that "the cyborg is our ontology," as it demonstrates (both in its practical integration with technology and as a revealing metaphor) the fundamental structure of being human (Haraway, 1991). Similarly, Dutch philosopher Maarten Coolen clarified that he too was "interested in precisely those anthropological ideas that one can associate with the act of technological transformation itself. What can man learn about himself from his own fabrications?" (Coolen, 1992, 165, 166; English translation by Peter-Paul Verbeek in De Mul, 2013, 234).

In a general sense, the transformation referred to by Coolen consists of experiences that elicit profound changes in the people that engage in them. The transformative quality of these experiences consists, according to Edmund O'Sullivan, of their triggering deep, structural shifts in the basic premises of our thought; they irreversibly alter the way we are in the world as subjects (O'Sullivan, 2003). Transformative experiences can emerge from partaking in activities designed to provoke certain psychological, behavioral, or convictional effects. In this sense, engaging in meditation exercises, going through psychological therapy sessions, completing a university course, keeping a personal diary, participating in team sports, and mulling over a thought-experiment are often considered to be emblematic transformative experiences[7] (Gualeni, 2014b).

Not all transformative experiences are, however, the product of transformative practices, and not all of them emerge from engaging in activities specifically intended to promote particular transformations. Transformative experiences can, in fact, also emerge from events and activities that were not specifically meant to elicit shifts in our self-understanding and conduct. Changes in our ethos and in our sense of possibility can be triggered by virtually anything – for instance, the appreciation of a specific piece of art or literary work, a hike in the Bavarian Alps, or a near-death experience.

This second group of transformative experiences (the accidental, subjective ones discussed in the previous paragraph) tends to come about in ways that are relatively rapid and unexpected, and that appear to elicit accidental and personal changes in their recipients. Differently from the experiences in the second group, those belonging to the first group (the ones specifically designed to elicit transformation intersubjectively) can be seen as having a more universal appeal and as emerging from a deliberate engagement in certain transformative practices. Additionally, the transformative experiences that derive from partaking in transformative practices customarily require an extended period of time for their transformative effects to take place (e.g., in the case of completing a university course or participating in therapy sessions). Some require a continuous or even lifelong engagement with the transformative practice in question (e.g., in the case of keeping a personal diary or meditating) in order for them to influence how we develop our sensitivity and shape ourselves and our ethos (Gualeni, 2014b).

While studying classical antiquity Michel Foucault encountered an ethical approach that was not primarily about showing or establishing which kinds of behaviors are to be considered morally right. Foucault argued that, in ancient Greece, in contrast to contemporary Western culture, the discipline of ethics did not aim to answer the question "How should I act to be a moral subject?" but rather "What kind of subject do I want to be?" Put more simply, in ancient Greece, ethics were not practiced normatively, but projectually, as a form of self-design. Methods and practices that were involved in experimenting with one's dealing with pleasures and knowledge, other people, political life, and styling oneself were labeled by Foucault as "technologies of the self" (Foucault, 1982; Verbeek, 2011, 75). He defined technologies of the self as practices that permit individuals to perform, alone or with the help of others, "a certain number of operations on their own bodies and souls, thoughts, conduct, and way of being, so as to transform themselves..." (Foucault, 1988, 18).

The reader will certainly recognize in Foucault's description what was earlier described as transformative practices. According to Foucault, when we engage institutions and networks of power through technologies of the self (or transformative practices), we become aware of the arbitrariness of institutions and "show which space of freedom we can still enjoy and how many changes can still be made" (Foucault, 1988, 11). In this sense, by reframing our understanding of power and freedom, Foucault prompts us to realize that we are actually freer than we think we are. It is important to note, however, that not all technologies of the self are automatically also practices of liberation. For Foucault, to exercise one's freedom as an ethical subject, the transformative practice in question needs to be framed in a certain way by the individual; for the practice to be liberating, one must first identify the constraints imposed by power that are shaping one's subjectivity and then deliberately engage and challenge the societal rules in question (Foucault, 1982). In other words, Foucault argued that it is only through a critical approach to one's constraints and interdictions that a technology of the self can also have a liberating effect on the individual. Foucault hoped that this particular way of self-fashioning, this critical and flexible engagement with power, could lead to the kind of freedom discussed above – the exerting of power over oneself in the same way that an artist exerts power over his tools and materials to produce a work of art (Parker, 2011).

Is the creation of a literary piece or a philosophical treatise, thus, as changing an experience for the writer as it is for the reader? And, aligning with the core concern of this section, are the designers of virtual worlds and simulated experiences going through a process of self-transformation while engaged in the process of designing their technologies and technological content?

I argue that, when engaging in the interrelated processes of framing projectual objectives, planning, creating, iterating, and evaluating that take place in the design of any experience (regardless of its simulated nature), the designers themselves acquire new knowledge. This new knowledge and the new perspectives developed by the designers of a simulated experience must be recognized as having a wide epistemological range, as it includes, at the very least:

- understanding the needs and the (perceptual, cognitive, and operational) capabilities of the intended recipients of the virtual world or simulated experience

- thoughtfully crafting the aesthetic metaphors that are present in the simulated world as well as its various interfaces and feedback mechanisms
- envisaging and controlling the effects of their design decisions through observation, grounded theory, and traditional quantitative methods
- recognizing their attitudes and beliefs in relation to the specific goals and ideologies embedded in the designed experience.

In any declinations and in any applications of the process of design, designers not only materially realize their functional plan, but also their ethos and their sensitivity, making these aspects of who they are and what they believe in objects for (the designers' own as well as other people's) critical evaluation (Gualeni, 2014b).

Heuristic, critical, and transformative aspects of the practice of designing experiences and interactions are not exclusive to the design of games and structured play, but are common in a critical approach to any form of design (Dunne and Raby, 2013). In the recent work of Dunne and Raby (2013), the design of artifacts, artworks, or experiences is recognized as critical when it directly addresses, challenges, and questions existing values and practices in culture, rather than adapting to them.

In the independent videogame development community, the idea of game design as a creative urge and an activity that is personally meaningful (beyond being an economic necessity) is well established and rooted in a tradition that dates back to bedroom coding in the 1980s. This aspect of the practice of game design is often discussed informally among game developers and is occasionally examined in specialized conferences and publications. Academia, however, has largely overlooked the idea that the design of virtual worlds can be engaged in by the designers as a method of self-reflection through technology, to clarify and realize their own beliefs and conduct, and to perform ethical and aesthetic self-fashioning in a practical fashion.

The conscious pursuit of game design and level design (or world-building) as practices of freedom is, I argue, more frequently and openly discussed among game developers in contexts such as serious games, games for change, self-reflexive videogames, and propaganda games. In other words, this aspect of game design is acknowledged in relation to applications of structured play that also have a deliberate transformative purpose for their players. Differently from the general and entertainment-oriented design goals of the games industry, in fact, the design of structured play as transformative practice always requires that their designers engage in 1) critical thinking, and 2) deliberately addressing actual socio-political questions and concerns.

In that respect, Italian activist and critical game designer Paolo Pedercini has often stated that, in his experience, "there is a greater liberation potential in designing games rather than playing games" (Pedercini, 2014). Pedercini creates videogames that have the overt scope of raising awareness and stimulating activism on certain socio-political issues. When designing critical games, according Pedercini, the foremost ethical responsibility of the game designer is to research and fully comprehend the topics and positions that they are simulating (procedurally and aesthetically) through their games (Pedercini, 2014). For example, if one were to design a game to call the attention of players to pathological gambling, it would be the designer's duty, at the beginning of the process, to develop a meticulous understanding of the chemical effects of gambling on the brain, of the multiple cognitive biases on which gambling relies, of the potential social harms that derive from gambling, of the possible solutions and policies that might be implemented to counteract these social harms, and so forth.

Later in the design process, the designers critically systematize their findings and concentrate on certain key aspects of the experience they are trying to disclose experientially and interactively. In doing so, designers customarily focus on the behaviors and qualities of the system they are designing that they consider to have the most rhetorical potential, and excise the ones that are superfluous or secondary and might muddle the clarity and persuasiveness of their message.[8] Jenova Chen, game designer of the award-winning, transformative 2012 videogame *Journey,* stated, "We build our games like a Japanese garden, where the design is perfect when you cannot remove anything else. I think that by doing that, the voice of your work is more coherent. If you have a lot of clutter on the top, the work may be more impressive, but you won't really know what it's trying to say" (Smith, 2012). These operations are normally carried out by the game designers in the pre-production phase of game development, but they are iteratively modified and refined throughout the entire process.

The transformative steps of researching, shaping an interactive experience, refining it, and making it clearly and easily accessible for the intended recipients are not exclusive to game design, but are characteristic of all forms of design engaged with the projectual ethics introduced above. Architecture, graphic design, industrial design, scenography, game design ...; in fact any form of design that permits the designers to take care of themselves – to critically confront existing structures of power and knowledge, and sociocultural limitations and interdictions, with the scope of shaping and fashioning one's ethos and beliefs – are potential contexts in which designers can engage in a liberating self-fashioning.

As observed by Foucault, the self-fashioning of an individual, one's "taking care of oneself" was, for the Greeks, the essential principle of cities and one of the main rules for social and personal conduct (Foucault, 1988, 19). The whole idea of the *polis* relied, for Foucault, on its members becoming conscious, ethical subjects. Socrates himself, when teaching people to occupy themselves with their beliefs and conduct, was actually teaching them how to take care of the city (Foucault, 1988, 20). There is a great and overlooked potential for personal transformation and the fashioning of better citizens (and generally more ethical and complete human beings) in the practice of design, when design is approached critically.

The challenges and processes of game design are analogous to those of other forms of design. Consequently, it would be unsound to argue a case for the remarkableness of videogame design based on some of its inherent qualities as a practice. However, I claim that videogame design has a particularly advantageous position when it comes to the self-fashioning of individuals. The pervasive penetration of computers into social practices, and the increasingly democratic and accessible nature of videogame development tools have made it faster and easier for everyone to distribute virtual worlds, discuss their playful and experiential qualities, and contribute to the development and the refinement of videogames, their worlds and their lore.

The speed and flexibility in the creation of videogames and/or simulated experiences is not, however, what affords videogame design a particularly interesting sociocultural role in the current historical setting. After all, we could also design and prototype critical experiences quickly and cheaply using pre-digital simulation tools (with paper, cards, pawns, cardboards, chalk, beads, etc.). From my perspective, the aspect that makes contemporary game design particularly desirable and relevant as a practice of liberation is that it affords designers the possibility of effortlessly releasing simulated experiences to a global community. A community of players–explorers–designers–makers that is willing to play, discuss, and participate in the refinement of an experience, and even, in some cases, to refashion it independently. In this sense, communities revolving around videogaming and videogame design have definite affinities with what Daniel Bell defined as a "caring society." In other words, the digital medium is recognized here as being particularly efficient in allowing designers to realize not only their functional plan, but also their ethos and their sensitivity, making them objects for (their own as well as other people's) critical evaluation. I claim that this particular advantage chiefly revolves around the volume of simulations, videogames, and design-related information that is exchanged globally; and that this advantage is already earning videogame design a place among the dominant technologies of the self of the twenty-first century.

I would like to conclude this chapter with a cautionary note borrowed from media theory that prompts us to always be wary. Different ways of establishing relationships with ourselves and with reality through mediators necessarily entail a balance between the increase in acuity of certain cognitive functions and the desensitization of others. As designers, we need to remain aware of the fact that the digital medium is not a neutral instrument in shaping ourselves and pursuing liberation; it inherently poses external constraints that are themselves shaping our design ideas, our expressive possibilities, and our freedom.

5

Augmented Ontologies and a Challenge to Western Philosophy: Videogames and Simulations as Mediators of Human Thought and Experience

The previous chapter discussed the possibility of employing the virtual worlds of computer simulations and videogames to influence human cognition at a basic, structural level. Even when directed toward more practical fields of application, such as ethics and self-discovery, the perspectives on the philosophy of technology offered in Chapter 4 examined the possibilities and effects of digital mediation in very abstract and almost entirely theoretical terms.

This chapter continues and deepens the discussion on the ontological shifts triggered and encouraged by experiencing digitally-mediated (and often unworldly) worlds, it also begins to integrate theoretical insights with examples and observations derived from the practice of videogame design. The analysis of three different videogames – philosophical videogames that I designed or with which I was otherwise creatively involved – provides examples and inspiration to discuss and understand what is it like to develop thought with the assistance of computer simulations, both as a creator of worlds and as a subject in those worlds.

In this chapter, the practice of "doing philosophy" will be proposed as a specific form of mediation of thought that is supplementary to the philosophical tradition of textual expression and could even, in extreme cases, constitute an alternative to it. Don Ihde presented a similar perspective when he wrote that "[w]ithout entering into the doing, the basic thrust and import of phenomenology is likely to be misunderstood at the least or missed at the most" (Ihde, 1986, 14). Embracing postphenomenology as a philosophical framework helps to reveal the practice of doing within digital simulations (in the sense of both creating virtual worlds and acting within them) as a novel and experiential branch of philosophy.

5.1 What is it like to be a (digital) bat?

In explaining the difficulties in the articulation of an objective physicalist approach to the philosophy of mind, Thomas Nagel argued in his 1974 essay "What Is it Like to Be a Bat?" that human subjectivity is inescapably confined within the experience of what it is like to be a human being. Nagel began his essay with the assumption that empirical observations provide the basic material from which human beings perform any cognitive process. Experience is not only presented as the fundamental substrate for the construction of ontologies, but also as the essential ground from which human beings may imagine and adopt alternative worldviews. Based on that postulation, Nagel maintained that it must be impossible to widen or alter human subjectivity by representational means. According to Nagel, the subjective imagination needed to make sense of representational media can only suggest what it would be like for a human subject to perceive and behave as a bat. This is precisely because the subjectivity of a bat is not presented immediately and objectively to human beings. Rather, people experience what it could be like to be a bat through the existing filters of the perceptual, cognitive, and operational structures that constitute human subjectivity. What it is like for a human subject to perceive and behave as a bat was not, however, Nagel's concern. He wanted, instead, to ascertain whether humans could ever be capable of knowing what it is like *for a bat* to be a bat.

As outlined above, Nagel claimed that only experiences that have the quality of being objective can be used in a physicalist model. As a consequence of this fundamental stance, as is already suggested by the question posed in the title of Nagel's essay, it appeared evident to the American philosopher that the answer he was looking for could not be found in the context of the capability of the human mind to abstract and fantasize. What Nagel believed, in very practical terms, is that, in trying to understand what it is like to be a bat, it does not help to:

> imagine to have webbing on one's arms, which enables one to fly around at dusk and dawn catching insects in one's mouth; that one has very poor vision, and perceives the surrounding world by a system

> of reflected high-frequency sound signals; and that one spends the day hanging upside down by one's feet in an attic. (Nagel, 1974)

Nagel further observed that the ways humans understand the world and relate to it have an unavoidably subjective character. According to Nagel, this subjectivity implies that no worldview can be objectively experienced in the truth of propositions describable in human language (Nagel, 1974). The impossibility of completing the objectification of alternative phenomenologies is not confined to perceptually alien cases, like those of a bat, a whale, or a mosquito; this impossibility is also commonly experienced between human beings.

Having recognized that the answer to his question could not be handled objectively within the limitations of native human kinds of ontologies and therefore could not be elaborated in human language, Nagel decided to conclude his essay with a speculative proposal that temporarily set aside the philosophically problematic relationship between the mind and the brain. He envisaged the hypothetical possibility of closing the gap between subjective and objective knowledge from a direction other than human imagination. What Nagel hypothetically proposed was the creation of an alternative phenomenology that was not based on imagination or subjective representations.[1]

> Though presumably it would not capture everything, its goal would be to describe, at least in part, the subjective character of experiences in a form comprehensible to beings incapable of having those experiences. (Nagel, 1974)

"What is it Like to Be a Bat?" was written before the social diffusion of computers. At the time, Nagel could not have anticipated the consequences and opportunities to be offered by the advent of a technology capable of materially disclosing interactive and persistent experiences of virtual worlds and of virtual alternatives to the self. Drawing on the previously introduced definition of reality (see Chapter 3), and observations on the difference between fictional and simulational media forms (see Chapter 4), virtual worlds are seen as capable of granting access to what must be considered effectively real experiences. Videogames are particularly obvious examples of how objective and yet unworldly worlds can be encountered and experienced through the mediation of computer simulations. *Miegakure*, for instance, is an experimental puzzle-platformer videogame designed by Marc ten Bosch that challenges players to actively solve puzzles in four spatial dimensions.[2] While characterized by a fairly classical ludological structure, *Miegakure* has the overt objective of offering its players the experience of an interactive reality that is obviously incongruent with their everyday, proximal experience of the world. Commenting on the unworldly qualities of its virtual world, award-winning game designer Jonathan Blow stated in an interview that *Miegakure* is

> a valuable contribution to human experience, right? ... The games I like are ones that have shown me something I wouldn't otherwise have seen, and Marc's creating an experience that would not have been possible to have, had he not made it. (Clark, 2012)

The *Independent Games Festival* 2010 student showcase entry *Haerfest*, a videogame developed under my supervision by Technically Finished (2009), is another example of a deliberately uncanny videoludic world. Similarly to *Miegakure*, *Haerfest* offers the interactive experience of a distinctively extraordinary phenomenology. *Haerfest* was developed as a digital formulation of the questions posed in Nagel's 1974 essay.[3] Both in Nagel's text and in our experimental videogame, the choice of a bat was motivated by the fact that it is a creature that is relatively close to the human animal from a phylogenetic point of view (mammal, chordate), and yet, at the same time, is endowed with cognitive equipment that is deeply dissimilar from that of human beings. It is for this reason that Nagel referred to being a bat as an example of "a profoundly inhuman subjectivity."

According to Nagel's view, there is no way of knowing or reproducing the real consciousness of a bat. What *Haerfest* attempts to do is to objectify part of the subjective character of what it is like to be a bat *for a bat*. The game allows the human player to experience having very limited eyesight, flying by flapping flabby wings, and perceiving volumes of information via the discontinuous input of a sonar system (see Figure 5.1). Although its correspondence to the experience of actual bats is unverifiable, the world experientially disclosed by *Haerfest* is decidedly incongruous with the ways human beings relate to the actual world in their everyday lives. Even though contemporary virtual technology cannot (or perhaps cannot yet) objectively reproduce the subjectivity of a bat, this technology does effectively offer ways of revealing previously inaccessible alternatives for how we can potentially relate to worlds. The crucial point in this understanding of the ontological relevance and cultural role of interactive digital mediation is that several aesthetic and logical aspects of the virtual worlds that they disclose simply could not be encountered or experienced by human beings in the ordinary relationships that they can establish with the world to which they are biologically native.

Nintendo's 2006 *Legend of Zelda: Twilight Princess* is a commercial videogame that famously disclosed a digital world through inhuman perceptual systems borrowed from the animal world as part of its gameplay. In that videogame, Link, the young and anthropomorphic playing character, can enter a magical dimension where he takes the form of a wolf. As a wolf, Link acquires the ability to follow olfactory trails, which players perceive spatially as trails of permanent, colored smoke (Nintendo EAD, 2006). *X-Men Origins: Wolverine* (2009) and *Batman: Arkham Asylum* (2009) also feature avatars whose beastly nature grants their players perceptual and physical abilities that transcend human ones (including echolocation, visualization of smell trails, thermal vision, etc.) (Raven Software, 2009; Rocksteady Studios, 2009). *Haerfest* and the videogames mentioned above share the characteristic of providing the player with sensory systems that are augmentations of, or synesthetic replacements for, those to which humans are biologically bound.

Figure 5.1 A screenshot of Technically Finished's 2009 videogame *Haerfest*, showing a first-person simulation of the combination of a bat's short eyesight and its echolocation system

This proposed understanding of virtual worlds as mediators of thought and experience does not align with the perspectives or the agenda of object-oriented philosophy, a contemporary subset of speculative realism that is characterized by the rejection of ontological anthropocentrism (or correlationism). In my work, I do not intend to suggest that human beings will ever be able to objectively disclose the phenomenology of bats, mosquitoes, plastic bags, or tacos, or to access the exotic experience of their (alleged) inner world. As explained in the introductory chapter, this book relies instead on a large and encompassing understanding of humanism as its fundamental context and therefore understands the ontological effects of interactive digital simulations as extensions, fragmentations, multiplications, and distortions of a family of ontologies that are distinctly and inescapably human. Any pretense of experiencing or understanding alien phenomenologies and object-oriented ontologies is not only outside the philosophical scope of this book, but also appears to be motivated by a form of anthropocentrism that is more naïve and arrogant than the correlationalistic one condemned by object-oriented philosophy. Likewise, Nagel argued in the conclusive passages of "What Is it Like to Be a Bat?" that humans have no way of objectifying an alien worldview and, as a consequence, no way of knowing what such experience is like within the frameworks of either phenomenology or neuroscience. Nagel further observed that, even if we do ever reproduce the perspective of a bat, a mosquito, or a whale objectively, such an experience would not be received by human subjects as anything like the animal's experience unless the human's biology were fundamentally altered (Nagel, 1974).

For the reasons advanced in "What Is it Like to Be a Bat?" it should be apparent that it is indefensible to state that *Haerfest* can provide an interactive phenomenological account of the experience of what it is for a bat to be a bat. *Haerfest* is patently a technological artifact; the materialization of a world designed by humans to be engaged by humans that is mediated by a machine characterized by logics that are simplifications, extensions, and distortions of certain aspects of human rationality. Accordingly, virtual worlds cannot be understood as technological artifacts capable of disclosing radically new phenomenological and ontological horizons. However, virtual worlds can be recognized as pragmatically opening up new and interactive horizons of thought, and of ways to understand time, space, properties, and causation that are supplementary, and in some cases even alternative, to those through which human beings structure their everyday relationships with the actual world (Gualeni, 2014a).

Virtual worlds, thus, extend, fragment, multiply, and distort traditional human kinds of ontologies. This claim, which is central to my argument, is not exclusively applicable to the artificial worlds disclosed by ludic and videoludic applications, but is rather a quality that is inherent in any simulated experience and particularly obvious when those simulated experiences are organized and disclosed by a digital medium (from writing with a text editor to operating a flight simulator).

5.2 How to philosophize with a digital hammer

This section expands on the *praxis* of designing virtual worlds and virtual experiences with philosophical scopes and themes. I illustrate and dissect the philosophical approaches and videogame design choices that defined two overtly philosophical videogames (*Gua-Le-Ni* and *Necessary Evil*) that – unlike *Haerfest* – I personally designed and developed with the intention of complementing my more conventional (textual) academic efforts. *Gua-Le-Ni* is a commercially released, action-puzzle videogame that I designed and developed in collaboration with the Italian developers Double Jungle S.a.s. for the Apple iPad and iPhone platforms between 2011 and 2012, and *Necessary Evil* is a free, self-reflexive videogame developed as a contribution to the panel "G|A|M|E on Games: the Meta-panel" at the 2013 DiGRA conference in Atlanta, Georgia, USA.

Gua-Le-Ni was originally designed with the intention of playfully negotiating Scottish philosopher David Hume's notion of complex ideas (see Figures 3.2 and 5.2). According to Hume, most people possess the mental concept of a Pegasus. This is patently due to our being exposed to Greek mythology through a number of different media forms. This idea is also ostensibly true in the present century, where the Pegasus can still be encountered in books, illustrations, and several other adaptations of its folklore. In general, the Pegasus is presented as a divine horse that could fly using its legendary eagle wings. In Hume's work, the mythical steed is used as a paradigm for something that humans cannot encounter in the world they share as biological creatures but that they can imagine and contemplate. Nobody can, I believe, truthfully claim to have seen a Pegasus, to have ridden, smelled, or touched it, and yet the Pegasus is an idea that humans can fantasize about, discuss, write legends about, and so on (Hume, 1738).

For Hume, the idea of a Pegasus does not fall into the category of simple ideas, those ideas that can be caused simply by immediate sensory impressions of worldly objects. The Pegasus must, therefore,

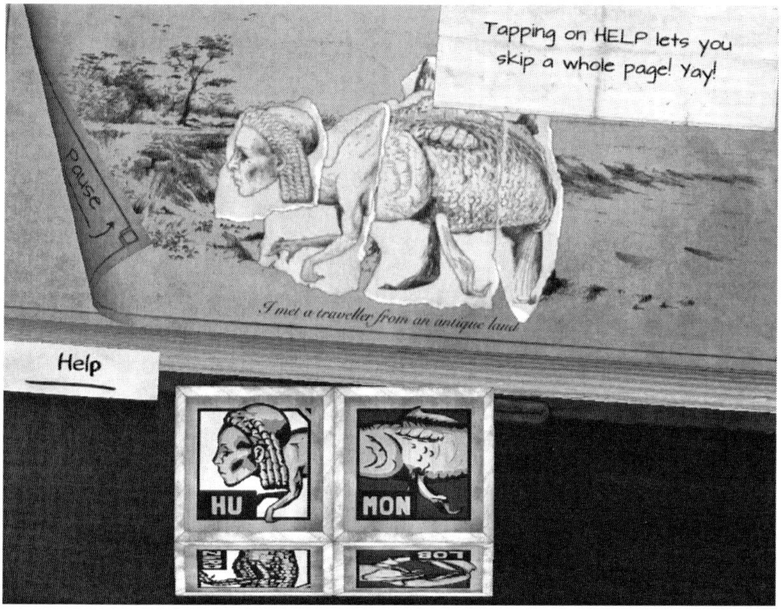

Figure 5.2 An additional game mode of *Gua-Le-Ni* also features (and trivializes) human body parts

be recognized as a complex idea, a creative, mental combination of elements and properties of which the human mind had a previous sensory experience. By means of fantastic beasts of the same combinatorial nature as Hume's Pegasus, *Gua-Le-Ni* prompts players to turn the creative capabilities described in Hume's *A Treatise of Human Nature* on their heads, and use them as elements of play. In *Gua-Le-Ni*, impossible paper beasts parade across the screen (the page of a fantastic bestiary) while the player attempts to identify the combinations of actual animal parts presented.

The Italian independent developers' community website (www.indievault.it) quoted a passage from a discussion on this point in which I explained that

> [i]f one learns how to play the game, one has implicitly understood Hume's essay, regardless of whether one aspired to do so or not. The player does not need to use her imagination or her interpretative capabilities in accessing those concepts of Hume's precisely because the game offers that portion of his thought in the form of an objectively

present, interactive allegory. (http://www.indievault.it/2011/11/23/gua-le-ni-una-perla-made-in-italy-per-ipad/ – translated from Italian)

Through both my videogames and my textual work, I argue that, by materializing philosophical perspectives, concepts, hypotheses, and thought, virtual worlds are influential in contributing to the rise of a new humanism and of new ways to pursue philosophical enquiry. My experimental videogame design titles demonstrate that the versatility and programmability of the digital platform can foster both the development of novel approaches to old philosophical problems and the creation of entirely new ones. On these premises, it is foreseeable that – facilitated by an increase in computer literacy, the growing accessibility of videogame development tools, and the progressive diffusion of digital media in social practices – more philosophical questions will arise and be confronted specifically within virtual worlds.

It must be specified, however, that I am not advocating the abandonment of written text, and I am not claiming that virtual worlds are (or are going to be) the ultimate philosophical mediators. What I am proposing in this book is an approach to the development of culture that can, where contextually desirable, hybridize or even replace traditional forms of media with simulational ones. In this sense, it is interesting to observe that while playing *Gua-Le-Ni*, when the game is in a paused state, the player is free to browse the pages of the virtual bestiary. Those pages contain randomly recombined paragraphs from articles and papers that I have published on the topic of the philosophical employment of games, in particular those explaining *Gua-Le-Ni* itself and its Humean inspiration. In this way, the philosophical message that motivated the videogame may be accessed (within the same virtual world) both through traditional media and within the simulation.

A second videogame that I developed as a practical demonstration and application of the philosophical methods discussed in this book is *Necessary Evil*, a self-reflexive videogame freely playable on any personal computer. The philosophical intuition that inspired *Necessary Evil* consisted of the realization that all the virtual worlds we are building do nothing else but objectify an idealistic perspective on reality. According to a radical version of idealism, the qualities that we can experientially encounter in objects (regardless of whether they are part of the actual world or a digitally-simulated one) are not objective properties; it is our experience of those objects – for example, in the case of George Berkeley's subjective idealism – that is responsible for bringing them and their properties into existence as mental contents (Gualeni, 2015).

The virtual worlds of simulations and videogames are customarily conceptualized and developed with the design goals of affording certain player-experiences and eliciting certain emotions through combinations of aesthetic stimuli, interaction, and narration. Similarly, from the specific perspective of software architecture, virtual worlds are technically structured around the user's (the player's) ability to perceive and interact with them. I believe it is revelatory, as an example, to reflect on the fact that in the customary, contemporary way that virtual worlds are technically materialized, objects that are too far away from a player, those whose sight is blocked by other objects, and those that are momentarily irrelevant for gameplay effectively do not exist as far as the game states are concerned. This particular way of structuring the experience of virtual worlds has the blatant functional purpose of limiting the number of calculations needed for a computer to materialize the virtual world suitably. Technically speaking, it is a desirable – if not necessary – evil.

Necessary Evil was designed with the intention of problematizing and demystifying the unquestioned idealistic structuring of videogames in a playful and interactive fashion. By doing so, it also inevitably ridicules the player-centrism of videogame worlds.[4] This purpose is pursued in the game design principally by having the player control a contributory character, a disposable minion of evil.[5] Players who are familiar with those conventional ludo-narrative structures that are well established in the videogames industry will recognize in the evil minion a generic, marginal character who customarily plays a secondary role in the progress of the main character. The main character of videogames is traditionally a (male, white) hero, whose courage and selflessness are meant to guide him on his mythical journey (see Figure 5.4). In *Necessary Evil*, and in sharp contrast to this tradition, the player controls a disposable baddie, while the hero is a computer-controlled non-player character (NPC). The horned minion of evil controlled by the player is confined in a dark cellar of sorts (see Figure 5.3). The minion is deprived of any consequential interactive possibilities with the room in which he is locked. This design decision was meant to make the player experience the feeling of marginality and to reveal experientially what a virtual world feels like when this world is designed around someone else's point of view and someone else's goals and desires. In the one room that the player can experience in *Necessary Evil*, there is nothing with which the player can meaningfully interact: doors do not open for the player-avatar; chests contain nothing; and objects in the room respond like flat theatrical props to the limited interactive possibilities of the red, horned monster.

Figure 5.3 In *Necessary Evil*, the player's interaction with the environment is entirely pointless; the little horned minion of evil controlled by the player cannot significantly interact with the room or escape from it

Figure 5.4 In *Necessary Evil*, the hero is an eloquent and relentless non-player character whose objective is to defeat monsters and vanquish evil; he attacks the monstrous player-character on sight

The game world is presented as existing for the unique scope of being experienced and traversed by the NPC-hero. The presence of the playing character (the horned minion) only serves as a challenge to the hero, an obstacle to be overcome to continue on his heroic quest. Once the

NPC-hero finally kills the little horned monster, he opens the door and leaves the room. At that point, the room and the player-creature are swiftly removed from the computer's memory, *de-rezzing* into nothingness and leaving behind a black, empty screen. The de-allocation of the game elements and their disappearance signifies the end of the experience for the player.

The first two sections of this chapter have analyzed three videogames that exemplify the possibility of utilizing virtual worlds for explorative, experimental, and didactic philosophical goals, demonstrating how videogames can be employed to communicate and negotiate philosophical ideas and hypotheses, materialize interactive alternatives to the *status quo*, implement thought experiments, and disclose new phenomenological possibilities. The first game, *Haerfest*, was an experiment in research-by-design that tried to reveal a world to players aesthetically, through the interactive experience of an alien phenomenology. The second game, *Gua-Le-Ni*, was designed to teach a philosophical notion implicitly through the practical activity of play, while its visual aspects play an ancillary, functional role in the experience. The third game, *Necessary Evil*, by presenting its critical worldview and philosophical arguments, both in the form of gameplay and through its aesthetics, can be understood as a combination of the previous two design strategies. In sum, the digital medium offers twenty-first century philosophers the opportunity to develop, test, and distribute their ideas in the form of interactive digital media content or with the added support of simulated worlds. I propose to call this new, experimental, and experiential branch of philosophy "augmented ontology" (Gualeni, 2014a).

5.3 Beyond the exclusivity of language: the philosophical practice of "doing"

Since Plato, the history of philosophy has been the history of written philosophy. This section elaborates on how the expressive and interactive possibilities of the digital medium can transcend many of the limiting effects that the traditional association of thought with text had on mental processes. Among the first scholars to advocate a critical attitude toward the exclusive and unquestioned association between thinking and writing was philosopher Ludwig Wittgenstein, who, according to Kristóf J Nyíri, was almost addicted to going to the movies and often used film to illustrate his philosophical position. Apart from a few remarkable exceptions, including Wittgenstein's *Tractatus Logico-Philosophicus* (1922) and Jacques Derrida's *Glas* (1974), philosophical books do not generally support their perspectives and arguments through their physical design and editorial structure.

Motivated by a comparable critical standpoint, philosopher and game designer Ian Bogost went as far as to accuse the exclusivity of the practice of writing of being a detrimental and unquestioned habit for the humanities in general. In his 2012 book, *Alien Phenomenology, or What It's Like to Be a Thing*, Bogost emphatically voiced his concern that "[t]he long-standing assumption that we relate to the world only through language is a particularly fetid, if still bafflingly popular, opinion" (Bogost, 2012, 90).

The material activity of doing philosophy, through which I propose to overcome the traditional association between the production of thought and the production of language, considers virtual worlds to be particularly interesting and engaging for mediating philosophical thought. This proposition has definite analogies with the concept of building (understood as an academic practice) in the connotation introduced by Davis Baird. In Baird's view, building – doing, constructing as a heuristic practice – offers an opportunity

> to correct the discursive and linguistic bias of the humanities. According to this view, we should be open to communicating scholarship through artifacts, whether digital or not. It implies that print is, indeed, ill equipped [sic] to deal with entire classes of knowledge that are presumably germane to humanistic inquiry. (Ramsay and Rockwell, 2012, 78)

In relation to the academic practice of doing, Bogost defined carpentry as the activity of constructing artifacts as a philosophical practice that "entails making things that explain how things make their world" (Bogost, 2012, 93). Two aspects shared by Baird's concept of heuristic building and Bogost's practice of carpentry are analogous to the philosophical approach to the mediation of thought proposed in this book:

1. their openness regarding non-textual options for the mediation of philosophical concepts, with their objections to the exclusivity of text and its largely unquestioned limitations on the activities of thinking and the dissemination of thought, and
2. their vision that the very crafting and framing of ideas and worldviews through a medium which is not necessarily concerned with the communication and production of semiotic meaning is in itself

a deeply philosophical activity and a transformative practice (as described in Section 4.3).

When proposing computer simulations as viable mediators to be employed in the pursuit of philosophical (or more widely intellectual) objectives, a line of reasoning frequently encountered to argue against this use of digital media contends that books are (and always will be) necessary and desirable. This opinion is supported by the observation that words afford the subtlety needed to symbolize and organize complicated arguments. According to the detractors of simulational media, this is something that virtual worlds cannot (or cannot yet) suitably materialize. As already contextualized earlier in this chapter, *Virtual Worlds as Philosophical Tools* does not advocate the abandonment of text in favor of artificial, interactive worlds, nor does it advance the claim that computer simulations are (or ever will be) the ultimate philosophical medium. However, it is my claim that there are no logical reasons why it would be generally ill-advised to embrace a vaster and more flexible media horizon with the intention of developing, expressing, testing, and divulging ideas.

Writing, after all, far from being a neutral way of exchanging information, has evident and inevitable constraining effects on the production of thought – effects that have been the focus of philosophical debate since the very introduction of writing in ancient Greek culture. A very strong argument against the primacy of a written approach to philosophy was, in fact, raised as early as the fourth century BC. In the Socratic dialogue, *Phaedrus*, Socrates brings to the attention of his Athenian interlocutor the various shortcomings of the written medium that was novel and controversial in Greece at that time. According to Socrates, written argumentation – unlike the dialogic development of a certain perspective or belief – cannot be adapted and shaped to meet the knowledge and the capabilities of the people to whom it is addressed, since it is a unidirectional form of mediation that is not open to compromise or negotiation.

Although videogames might not be suitable for presenting abstract concepts in their full intricacy and subtlety, traditional books offer the reader neither agency nor the possibility of negotiating with the objectified thoughts that the books mediate. If we exclude the choice of whether to continue reading, linear books – like any traditional media – only allow hermeneutical forms of freedom. In addition, as forms of mediation, books cannot embed objective representations of spatial contexts, whereas digital simulations can materialize spaces interactively and with relative accuracy (Gualeni, 2014a).

I would like to conclude my proposition for augmented ontology as a novel and digitally-mediated philosophical context with a cautionary remark borrowed from the field of media philosophy. In his seminal 1964 book, *Understanding Media: The Extensions of Man*, Marshall McLuhan observed that technology not only provides advantageous enhancements of human mental and bodily capabilities, it is also a form of self-amputation. In other words, new ways of establishing relationships with reality through media necessarily entail a balance between the increase in acuity of certain cognitive functions and the desensitization of others (McLuhan, 1964). With these effects in mind, the embedding of videogames and computer simulations in social practices (philosophy being one of them) might best be pursued with the awareness that videogames, like any form of mediation, can disclose reality in specific ways that are always inherently both revealing and concealing.

5.4 The problem with philosophical "play"

Approaches to the design and theoretical understanding of virtual worlds that focus primarily on their affording some form of doing are common. From the artistic perspective on game design customarily referred to as proceduralism, for instance, the ways in which virtual worlds allow for the emergence of meaningful interactive experiences have their foundation in the logical structuring of their interactivity. In the specific context of games and videogames, the logical cores of the experience are commonly referred to as game mechanics. From this standpoint, virtual worlds are mechanically devised by game designers and are considered capable of establishing unequivocal, interactive relationships with their players. In other words, for the proceduralists, digital simulations and videogames can cause predictable effects on the cognition and behavior of the players. This is the ideological foundation upon which, according to the proceduralists, virtual worlds can be understood as viable media for delivering information, funneling behavior, and effectively functioning as persuasive technologies.

In his 2011 article "Against Procedurality," Miguel Sicart explicitly discerned that the allure of proceduralism "comes from its quasi-scientific discourse, from its efficient, postmodern argument that ties technology, systems and reason together, justifying the existence of games as a serious medium for expression" (Sicart, 2011). However, the proceduralist understanding of the experience of virtual worlds, and of play, can be criticized (and indeed has been criticized) as an incomplete and impoverishing depiction of what is, rather, a fundamental and irreducible activity (Sicart, 2011). According to the detractors of proceduralism, a valid and thorough understanding of play ought to be embraced in all its complexity, ambiguity, and expressivity. The approach generally presented by the proceduralists is restrictively focused on comprehending and predicting quantifiable and performance-oriented dimensions of play, while ignoring the freely creative, ritual, social, and transformative aspects that Bernard DeKoven identified as the "myth domain" of play (DeKoven, 2002). In other words, proceduralism is criticized for disregarding the ways of engaging with virtual worlds that are subjective, informal, and not strictly deterministic.

Additionally, as Ihde has noted, no form of technical mediation can establish a stable and completely predictable relationship with its users. According to Ihde, the sociocultural consequences of the use of any technology can never be solely determined by the (sometimes clumsily pursued) intentions of the designers. Rather, technologies are multistable; they are always appropriated and interpreted contextually by their users (Ihde, 1990). Further, unexpected behaviors and effects arise from unforeseen malfunctions of the artifacts that mediate human actions and decisions (Verbeek, 2011, 97–99).

The multistable qualities of technology often appear to me, both as a videogame designer and as an avid gamer, to be radicalized in our dealings with virtual worlds as creators and players. Unexpected behaviors, technical glitches, and events that were not anticipated by the designers are, in fact, commonly experienced occurrences in virtual worlds of all kinds. I believe this to be the case because virtual worlds in general (and the worlds of videogames in particular) are characterized by several types and levels of interaction that often intricately overlap. Given the focus of these worlds on behavior and interactivity, the complex and interconnected systems that support virtual worlds also need to afford a certain flexibility and expressiveness in their use. The autonomy granted to the players in those worlds often leads to behaviors and interactive possibilities that can potentially subvert and trivialize both the experiential goals and the semiotic meanings originally intended by the designers (Gualeni, 2015). The number of erratic and hilarious videogame glitches published daily on video-sharing websites is a testament to the imperfect control that game designers and developers have over their technological instruments.

An example of the awareness of the particularly penetrating multistability of virtual worlds was voiced in a recent interview for the *New Statesman* by Jason Rohrer, the independent author of celebrated experimental videogame titles such as *Passage* (2007) and *The Castle Doctrine*

(2013), and Merritt Kopas, designer and creator of *Lim* (2012), a free, web-based videogame about the tension of trying to meet society's expectations:

> "I think that systems have a tendency to get away from us," says Kopas. "We intend to portray or produce one thing, but the systems we're creating seem to resist or reshape our intents." Even Rohrer, with years of programming experience (this game is his seventeenth), has to take responsibility when things go wrong. "As a designer, I'm trying to build the tightest system that I can build. I don't want there to be those system leaks which allow bizarre readings, and involve the procedural rhetoric effectively falling off the rails and going who knows where." (Brindle, 2013)

If the possibilities for autonomous agency and the emergence of unexpected behaviors in virtual worlds threaten to distort and trivialize the affordances and messages originally set up by game designers, how could such worlds ever be treated as a medium for communication? How could a defined meaning ever emerge from contents that are not only infinitely interpretable (as was already the case for text and all other traditional media forms), but are also fallible and infinitely manipulable?

When embracing perspectives on acting in virtual worlds and experiencing virtual worlds that are broader and more encompassing than those of the proceduralists, the job of game developers can no longer be seen as designing an experience. Rather, the role of game developers must been seen as contributory to the experience in the sense that they set the stage for experience to emerge (Salen & Zimmerman, 2003, 168). Abandoning a formal and deterministic understanding of virtual worlds and their effects, the figure of the creator of virtual worlds can no longer be associated with that of a demiurge, a divinity capable of creating worlds and controlling the fate of their inhabitants. This figure is instead identifiable with an earthly scenographer, someone who is responsible for setting up constraints and affordances that will be freely appropriated by the actors (the players) during play[6] (Gualeni, 2015).

It is my belief that neither the recognition of limitations to controlling messages and experiences in virtual worlds nor discontent with proceduralist approaches to play should encourage game scholars, game designers, and media philosophers to discard their insights and deterministic methods. The uncompromising rejection of scientistic ways of understanding play (understood both as an activity and as its experiential outcome) is in fact no less impoverishing than the excision of the freer and more subjective ways of understanding performed by procedurality. What I propose is, instead, to embrace deterministic approaches for framing our experiences in virtual worlds as instruments that are useful and revealing in specific contexts. Perspectives like ludology, procedurality, and game-user research can be usefully employed to uncover some aspects of the functional behaviors of simulations, and should be recognized as capable of helping designers and academics alike to anticipate and control some of the effects that design choices will elicit for the players. As observed by Sicart, the deterministic framework offered by the proceduralist approach can be fruitfully applied to analyze single-player videogames, or in general games that offer limited operative options to their players. The worlds of those videogames are, by their very design, structurally efficient in constraining player behavior, allowing the execution of a few restricted actions in the specific and limited ways envisaged by the game developers (Sicart, 2011). Among the videogame genres that funnel player behavior more starkly are those defined by a few player-related mechanics, such as puzzle games, simple resource management games, point-and-click adventures, 2-D platform games, and hidden object games.

The last two philosophical videogames discussed in this chapter were single-player experiences explicitly designed to direct player behavior toward simple and non-negotiable objectives by offering very limited operative options. As playful virtual worlds designed to restrict and funnel player behavior, both *Gua-Le-Ni* and *Necessary Evil* can be considered capable of explaining philosophical notions and articulating arguments in ways that are largely unambiguous. It must be noted, however, that – at least in principle – it should always be possible to develop interactive simulations and videogames with philosophical scopes and themes that are less constraining and more expressive than those purposefully designed to control play and to materialize a specific set of ideas. The interactive experiences of virtual worlds that allow for freer and more ambiguous types of play cannot lead to the emergence of a univocal and clear meaning but can still interactively disclose worlds that are alternative to those human beings experience in their everyday engagement with the actual world (as was the case with *Haerfest*, for instance). Put succinctly, all videogames allow their players to experience alternative phenomenologies, but not all videogames function as communication instruments.

I argue that the proverbial baby be saved from being thrown away with the dirty bathwater by means of a cautious and instrumental use of quantitative methods for understanding virtual worlds, both as designers and as scholars. In other words, formal and objective approaches to the analysis of our experience of virtual worlds and their effects on the players can be fruitful methods of obtaining a closer understanding of the sociocultural (and even political) impact of digital mediation. However, I claim that such methods will only be fruitful when employed with an awareness that our experiences and behaviors in virtual worlds remain complex and irreducible activities that cannot be completely anticipated or fully captured by questionnaires, interviews, or statistical analysis.

6
Positionality in the Digital Age: Virtual Bodies and the Effects of Virtual Experiences

6.1 General introduction to Chapters 6 and 7

Chapters 6 and 7 complete my postphenomenological perspective on the philosophy of (digital) technology. Chapter 6 articulates an understanding of the experiences of alternative, virtual worlds disclosed by digital simulations. This understanding will follow a chiefly anthropological standpoint, laying the groundwork for answering the question at the core of this book from the human side of the human–computer relationship. Chapter 7 (Virtual Worlds as Poetic Allegories) will then shift the focus of this inquiry closer to the technological side of the relationship. In other words, Chapter 6 elaborates on how the digital medium can fragment and overcome traditional human kinds of ontologies, and Chapter 7 will explore to what extent human ontologies can be distorted, multiplied, and extended in virtual worlds.

The separation of these aspects into two distinct chapters could be (mistakenly) interpreted as reiterating the Cartesian schism between the subject and object of inquiry. By subdividing my work in this fashion, I do not intend to conclude this book with a dualistic bifurcation. The arguments presented in this text build, rather, upon the belief that neither human ontologies nor virtual worlds can be thoroughly explored and rigorously understood independently of one another. This foundational stance was outlined in the introductory chapter through the presentation of the adopted postphenomenological approach to the philosophy of technology. In the engagement between mankind and the technological environment, specific objectivities (worlds) arise, as do specific subjectivities of human beings. Consequently, subjectivity and objectivity are not presented here as two independent poles between which an ontological relationship takes place. Rather, subjectivity and objectivity are considered products of their mutually-determining relationship. As such, they cannot be absolutized or fully comprehended outside the context of that relationship. Similarly, the textual separation between Chapters 6 and 7 is not meant to reflect a conceptual polarization or a dichotomy, but simply to provide the reader with a more thematically coherent presentation and an organization of the content that is easier to follow.

6.2 The relationship between technology and human thought: historical foundations

The ability to manipulate objects is a sophisticated behavior that is highly evolved in all kinds of primates (Umiltà et al., 2008). Due to this crucial evolutionary achievement, primates are able to interact with objects not only by using their natural effectors (hands, arms, teeth, etc.), but also by means of non-body tools. Recent experiments on monkeys revealed that, neurologically, basic utensils like sticks, stones, and small rakes become – with use – part of the bodily equipment of the test subjects (Iriki, Tanaka, Iwamura, 1996; Umiltà et al., 2008). As explained by Umiltà and colleagues in the 2008 paper "When pliers became fingers in the monkey motor system," both normal and reversed pliers become equivalent, with practice, to natural effectors as far as the acting body-schema of the monkey is concerned (Umiltà et al., 2008).

In the first half of the last century, Merleau-Ponty proposed an understanding of the relationship beings establish with artificial effectors through their bodies that is comparable to the neurobiological link presented above. This understanding was part of Merleau-Ponty's theory of embodied relations as an interpretative framework that was built on observations concerning how people engage with devices and other beings in the world. Characteristically deriving his examples from everyday life, Merleau-Ponty observed that:

> [a] woman may, without any calculation, keep a safe distance between the feather in her hat and things which might break if off. She feels where the feather is just as we feel where our hand is. If I am in the habit of driving a car, I enter a narrow opening and see that I can 'get through' without comparing the width of the opening with that of the wings, just as I go through a doorway without checking the width of the doorway against that of my body. The hat and the car have ceased to be objects with a size and volume which is established by comparison with other objects. They have become potentialities

of volume, the demand for a certain amount of free space. (Merleau-Ponty, 1962, 143)

More than a century before Merleau-Ponty published his theory of embodied relations, Danish inventor Hans Rasmus Johann Malling-Hansen created the world's first commercially produced typewriter, the Malling-Hansen writing ball. According to literary sources, Friedrich Nietzsche ordered a writing ball for himself. Having never fully recovered from injuries he suffered in his early twenties, with his health worsening by the day and at his wit's end, Nietzsche thought the writing ball would help him to resume his writing activities (Emden, 2005, 27–29; Carr, 2010, 18, 19). Upon Nietzsche's adoption of the writing ball, one of his closest friends, the writer and composer Heinrich Köselitz, noticed a change in the philosopher's writing style. "Perhaps you will through this instrument even take to a new idiom," Köselitz commented in a letter to Nietzsche dated February 19, 1882, also adding, regarding his own work, that "my thoughts in music and language often depend on the quality of pen and paper." "You are right," Nietzsche replied, "our writing equipment takes part in the forming of our thoughts" (Nyíri, 1993).

A remarkably similar observation to Nietzsche's and Köselitz's was later offered by Merleau-Ponty himself. Reflecting on the relationships that humans habitually establish with tools (and also using a typewriter as an example), he commented that "the subject who learns to type incorporates the key-bank space into his bodily space" (Merleau-Ponty, 1962, 145). According to the theoretical perspectives proposed by Merleau-Ponty, any kind of technical tool and bodily extension (temporarily) enhances the physical capabilities of mankind and the aptitude for manipulating things.

The initial recognition of the decisive influence of technology on the possibilities for humans to collect, rationally organize, objectify, transfer, combine, and preserve information dates a few centuries before neurological experiments on monkeys, before Merleau-Ponty's feathered hats, and even before the decline of Nietzsche's career as an author. It can be traced back to the very dawning of written philosophy, to Plato's writing of the Socratic dialogue *Phaedrus*.

In Plato's *Phaedrus*, Socrates and his Athenian interlocutor spend a torrid afternoon outside the Greek capital. They take advantage of a quiet spot under a shady tree to rest and wait until the midday heat becomes more tolerable while discussing the fine points of rhetoric. Phaedrus and Socrates begin to discuss the primary qualities of speech-making, but soon their dialogue veers to topics such as the nature of desire and the journey of the immortal soul. Toward its conclusion, the dialogue turns to the impact of the technically-mediated practice of writing on how people develop thought and pursue truth and beauty (Plato, 1995, 79, 80). At the time when *Phaedrus* was written, writing was a freshly introduced and controversial technique for storing and transmitting information. Unlike Plato (the author of the dialogue), the character of Socrates in the dialogue itself is resolutely opposed to the adoption of writing and is dubious about the consequences it could have on the production and circulation of thought and, ultimately, on Greek society. In the conclusive part of *Phaedrus*, Socrates openly argues against the transition of Greek culture from its traditional oral form to a literary one reliant on the externalization of thought and memory in the form of arbitrary symbols.

Explaining his argument, Socrates recounts the mythical anecdote of the encounter between Thamus, one of the ancient Kings of Egypt, and Theuth, the god whose many creations included the alphabet, according to ancient Egyptian mythology. In the story, Theuth presents the invention of the written word to Thamus and tries to convince the king of Egypt that his latest divine creation "will make the Egyptians wiser and will improve their memory" (Plato, 1995, 79). King Thamus, however, is skeptical and raises the concern that writing might have the very opposite effect, as it could "introduce forgetfulness in the souls of those who learn it: they will put their trust in writing, which is external and depends on signs that belong to others, instead of trying to remember from the inside, completely on their own" (Plato, 1995, 79).

Socrates then acknowledges that there are practical benefits in the framing of thought within a system of material symbols. He believes that writing can indeed lead to obvious advantages and a higher efficiency in certain contexts and practices, listing, for instance, the cases of bookkeeping, accountancy, and preserving memories from the forgetfulness that comes with old age. However, vicariously speaking through Thamus, Socrates contends that writing entails a progressive dependence of the mind on technological artifacts that will inevitably alter the way people make use of their minds. Socrates believes that substituting external signs for internal memories and sensations will only supply people with a shallow semblance of knowledge, preventing them from achieving the intellectual depth that leads to true knowledge and happiness.

In line with Socrates's beliefs, one of the foundational assumptions of media philosophy postulates that the mediated contents cannot be understood separately from how they are mediated. Correspondingly, historian and philosopher Walter J Ong has argued that the historical transition from oral cultures to cultures characterized by literacy led to a fundamental shift in how people developed and structured their thought. Both Ong and Eric Havelock went as far as to recognize Plato's Doctrine of Ideas as a direct consequence of the cultural adoption of the textual medium in ancient Greece (Ong, 1982; Postman, 2005).

According to the perspectives proposed by thinkers such as the later Nietzsche (post-writing ball) and Ong, the very emergence of theoretical thought in ancient Greece is specifically suited to being interpreted as a by-product of the externalization of mnemonic and thinking processes entailed in the adoption of phonetic writing. Ong specified that:

> By separating the knower from the known, writing makes possible increasingly articulate introspectivity, opening the psyche as never before not only to the external objective world quite distinct from itself but also to the interior self against whom the objective world is set. (Ong, 1982, 105)

One of the most striking functional advantages of recording and producing thought in written form over the preceding oral cultures lies in the fact that knowledge, which was necessary to survival and fundamental to the promotion of cultural values, was no longer bound to the imperfect and constraining biological system of recording, but could be objectified, duplicated, and consulted with better fidelity and on a more durable and reliable medium.[1]

As already observed, establishing a relationship with reality through media necessarily entails a balance between the increase in acuity of certain cognitive functions and the desensitization of others (McLuhan, 1964). While objectifying and extending the capabilities of human beings for preserving, communicating, and organizing thought, textual mediation can also inherently render the process of developing and communicating knowledge less personal and less flexible than in the oral tradition. Albeit open to interpretation, text does not afford the dialectical negotiation of meaning with the mediated information that was characteristic of oral cultures.

As previously noted, the introduction of any technical mediation into social practices and processes might be better pursued with the awareness that they disclose reality in ways that are both revealing and concealing. In this sense, the invention of writing is understood as having initiated a profound transformation in Western culture that had momentous repercussions from both a cognitive and an anthropological standpoint.

However, its foundational influence on Western thought notwithstanding, writing is not the only technology that had (and still has) a cultural, ontological, and even evolutionary influence on mankind. All technologies, in fact, externalize and extend certain human possibilities and facilitate the transportation and manipulation of beings, resources, and information. As such, they co-determine the shaping of new worldviews and disclose broader interpretative horizons to understand what it is like to be in the world as humans in certain historical contexts.

6.3 On science fiction and the bodily origins of human kinds of ontologies

> She could see the image of her son, who lived on the other side of the earth and he could see her...'What is it, dearest boy?'...'I want you to come and see me.' 'But I can see you!' she exclaimed. 'What more do you want?'... 'I see something like you... But do not see you. I hear something like you through this phone, but I do not hear you.' The imponderable bloom, declared by discredited philosophy to be the actual essence of intercourse was ignored by the machine. (Forster, 1985 in Goldberg, 2000, 48)

Inspired by Forster's 1909 pioneering techno-pessimistic novel, *The Machine Stops*, (already mentioned in the introductory chapter), a remarkable portion of the science fiction produced in the last century directed its concerned attention to what Hubert Dreyfus identified as our technically-induced "progressive loss of touch with reality" (Dreyfus, 2000, 48–63). The work of Philip K Dick is a particularly conspicuous example of such a vision. Socio-political, ethical, and fundamentally ontological themes are treated, in the body of Dick's literary work, as intimately bound to the potential of technology for simulating, integrating, distorting, and even replacing the world. This particular understanding of the interface between technology and human beings is pivotal in several of Dick's best-selling novels and short stories, including *The Three Stigmata of Palmer Eldritch* (1965), *Do Androids Dream of Electric Sheep* (1968), *UBIK* (1969), and *We Can Build You* (1977).

In science fiction, Dreyfus maintained, the recurrence of themes concerned with the progressive detachment of mankind from a direct and genuine relationship with the world leading to a complete refashioning or replacement of reality itself is not a mere fantasy, but a vivid metaphor for the way epistemology developed in the Western world (Dreyfus in Goldberg, 2000, 50). Dreyfus referred to the foundational dualistic stance that shaped Western thought and guided its growth up until now as a theoretical form of scientism.

As summarized in the first chapter of this book, in the seventeenth century Descartes posited a methodological distinction between the immaterial *res cogitans* (thinking thing) and the material *res extensa* (extended thing). This fundamental dichotomy divides the mental content of human beings (understood as independent from the perceptions of the world) from things that have material extensions (such as inanimate objects or our human bodies). In line with his methodological approach, Descartes defined the human mind as an immaterial substance that, differently from material objects, is *timeless*. Following the Platonic and Christian tradition, Descartes, further identified the *res cogitans* with the immortal soul.[2]

In his writings, Descartes never denied that our mental content depends in several ways upon our sensory equipment, various tools, and artificial extensions. According to Descartes, at least part of the way we organize and develop knowledge is always directly involved with the physical world. However, from his theoretical standpoint, the human mind could also exist independently of the physical world. Throughout the history of modern philosophy the ancillary role that Descartes assigned to sensory experience in his epistemological framework and the Cartesian disregard for the body-related foundations to the structuring of ontologies (or the production of any mental content for that matter) expectably met a strong opposition from thinkers embracing empiricist and materialistic perspectives. The latter argued that knowledge and the very process of thinking could not take place without a physical substrate. A famous adage by Jacob Moleschott asserts the impossibility of the emergence of subjectivity without a material basis to support it: "without phosphorus there would be no thought in the first place" (English translation in De Mul, 2010, 167).

In the twentieth century, the work of phenomenologists (such as Heidegger, Plessner, and Merleau-Ponty), pragmatists (like William James and John Dewey), and, so-called, philosophers of ordinary language (like John L Austin and the later Ludwig Wittgenstein) problematized and countered Descartes' dualism with alternative approaches. Offering each their own specific reasons and arguments, the branches and figures of modern philosophy mentioned above claimed that, among other methodological concerns, dualism was incompatible with our everyday experience of the world. More recently, an interesting critical position against a dualistic approach in philosophy of mind was offered by proponents of the extended mind approach, such as Andy Clark and

David Chalmers. Their work presents a form of externalism that this book, for reasons that will be elaborated upon later in this chapter, is very sympathetic with. They argue that our world and our bodily relationships with it have an active role in the shaping of our cognition and interaction; they function as parts of our minds. From their standpoint, theoretical separations between the mind, the body, and the environment preclude the possibility of developing a thorough account of what the mind is and how it works.

In a way that is analogous to the externalist take on philosophy of mind outlined above, this book understands the virtual worlds disclosed by digital simulations and videogames to be coessential to the development of augmented forms of human cognition, perception and agency. *Virtual Worlds as Philosophical Tools* is, therefore, concerned with areas of contemporary philosophy of technology where some of the skeptical positions upheld by Descartes' dualistic ontology seem to be well motivated. Especially in its simulational applications, the digital medium appears to be capable of taking the cunningly persuasive (if not openly deceptive) role of Descartes' evil genius. Confronted with increasingly sophisticated and easily accessible simulated worlds, it is to be expected that human beings might come to think of their being-in-the-actual-world as a specific, relative instance of a more encompassing way of relating to reality, rather than recognizing in their corporeality the essential grounding of their worldviews. Embracing this standpoint, the social diffusion of interactive digital media is likely to stimulate the re-emergence of cultural debate on questions about the authenticity of human experience, possibly leading – as was the case in the seventeenth century – to the rise and adoption of purely rational epistemological stances. Supported by the encompassing rhetorical potential of digitally simulated worlds, human beings could be persuaded to understand themselves as processors in vats, and to relate to their sensory systems as if they were data encoders.

In the age of digital mediation (and especially when trying to map concepts like agency, identity, and presence in virtual worlds), the most salient critiques of dualistic, theoretical epistemologies still largely elaborate on perspectives that were originally presented by Merleau-Ponty. In his influential 1945 book *Phenomenology of Perception*, Merleau-Ponty carefully presented objections against both empiricist and rationalist theories on human perception and agency. One of his central claims is what he called the "primacy of perception," the conviction that the perceptual and interactive bodily involvement with the world precedes, in human cognition, the possibility of theorizing about the world itself.

Merleau-Ponty believed that when our everyday engagement in the world proceeds as expected, we do not perceive ourselves (or theorize about ourselves) as subjects who relate to objects in the external world while having inner experiences. He believed, instead, that our relationship with the world stems from a fundamental bodily involvement that, with use and experience, becomes automatic to a degree and recedes cognitively into unobtrusive familiarity. This perspective was inspired by Heidegger's notion of something being ready-to-hand (something with which human beings can establish an immediate relationship and intuitively interact, and something whose use cognitively precedes the possibility of a theoretical, objective approach). Consequently, for Merleau-Ponty, the established and embodied involvement of human beings in the world is understood as a pre-intellectual activity.

From this phenomenology-inspired perspective, the subject and object of observation are not understood as independent ontological absolutes, as they would be understood in a dualistic approach; rather, they are embraced as mutually interdependent aspects of what it is like to be in the world as humans. They become two components of the same process, components that are inextricably tied together in the human body. Maarten Coolen noted, however, that in Merleau-Ponty's phenomenology, "the body itself seems to 'vanish' when it perceives something or puts itself into action, in favour of the world that is opened by it. It gets, so to speak, swallowed up in being attuned to the world" (Coolen, 2013, 66). Coolen claimed that, to be complete, a phenomenological account of perception and action must also take into account the capability of human beings to reflect on their own corporeality and on the engagement possibilities offered by their bodies. According to Coolen, in fact, it is this self-reflexive quality of human beings that allows them to establish a cognitive distance from the world they inhabit in the first place and that grants them the possibility of distinguishing the objective features in their surroundings (Coolen, 2013, 65).

The actualization of the self-reflexive relationship mentioned above was first recognized by German philosopher Helmuth Plessner as being particularly definitive of the way that human beings are in the world. The eccentric qualities of the human being and the consequent capability for self-reflection have a central role in the development of Plessner's philosophical anthropology.[3] Relying on Plessner's work, and in particular on his theory of positionality, or rather on De Mul's re-elaboration of this theory in the digital age, I structure an understanding of virtual worlds as mediators capable of enhancing and expanding the native body-schemas of human beings with supplementary, virtual body-schemas. Framing perceptual, cognitive, critical, and operational effects of the experience of virtual worlds on our fundamental, bodily ontological tools will be instrumental to articulating my answers to the questions concerning digital technology.

6.4 Human eccentricity and virtual homelessness

Plessner famously referred to the indeterminate and unfathomable nature of the human condition as "an open question." According to Plessner, the human being arises from a situation of uncertainty: "[i]n this relation of uncertainty with regard to himself, man comprehends himself as a power and discovers himself for his life, in theory and practice, as an open question" (Plessner, 1980–1984, Vol. IV, 321, English translation in Safranski, 2002, 206).

The ambiguous and unknowable anthropological constitution of human existence is elaborated in a specifically vivid and thorough fashion in Plessner's 1928 book, *Die Stufen des Organischen und der Mensch* (*The Levels of the Organic and Man*), in the form of three fundamental anthropological laws. Plessner's anthropological laws present man as an incomplete creature, an open project whose self-reflexivity inherently leads to aspirations, such as (1) reaching a state of completeness and satisfaction (or balance) that man innately pursues by means of culture and technology, and (2) cherishing dreams of "home," of finding a reliable ground on which to establish stable values and construct one's existence (Plessner, 2006; De Mul, 2010, 204–205).

"In this way, Plessner radicalizes the philosophical anthropological theme of man as a *deficient* being" (Verbeek, 2013, 236). Famously defining man as the "being that needs to be surpassed," Nietzsche accordingly presented being human as an ultimately incompletable task; a challenge in which mankind is constantly asked to overcome itself. Observing the dynamism and uncertainty that emerged from a crucially self-reflective existence, Plessner understood the human condition as being defined by not being definitively definable. Every ethical, scientific, or religious frame of reference for a possible definition of man is, according to Plessner, inevitably a historical product of man himself. Mankind, understood in this way, is always both the producer and the product of culture (Safranski, 2002, 206). "First, man made the hammer, and then the hammer made the man," as condensed in an aphorism attributed to McLuhan.

For Plessner, the characteristic openness and indeterminacy of the human condition also constitutes the foundation of political power. He understood the defining indeterminacy of the human condition as a fundamental requirement for human beings to express their freedom and to bring about sociocultural changes. Similarly, philosophers such as Heidegger and Flusser elaborated significant portions of their thought on (or rather toward) an understanding of man as an unfinished project and a groundless being. According to Heidegger in particular, *Dasein* (human existence, literally being-there) can only be fully comprehended as a project that shapes, and is relentlessly shaped by, his world against the backdrop of its own finitude.

For a long time, Plessner's work remained relatively unknown to the English-speaking world, obscured by the long shadows cast by his contemporaries (in particular, Max Scheler, Plessner's mentor and a central figure of philosophical anthropology, and Heidegger, whose *Being and Time* was published just one year before Plessner's *The Levels of the Organic and Man*). Only recently has Plessner's work started to gain academic momentum, mainly due to the application of his theory of positionality to the study of information and communication technologies. *The Levels of the Organic and Man* offers an insightful anthropological framework for understanding the relationship between our biological bodies and their extensions in virtual or telepresent environments. I argue that Plessner's work can complement the more widely read and discussed classical phenomenological frameworks and can be used as a starting point for structuring a rich and encompassing interpretation of embodied cognitive processes in the age of simulational media.

Plessner understood the body as the boundary of the organism, and as much part of the internal world as it is of the external (Plessner, 2006, 82). Such a boundary is intuitively perceived as the spatial limit through which a being encounters other beings (Plessner, 2006, 127). Plessner identified the relationship between a living organism and its boundary as the organism's "positionality." In Plessner's philosophical biology, positionality defines the spatial structure and cognitive and operational autonomy of an organism in relation to its instincts and its surrounding environment. The spatial structure of the positionality of an organism is what, according to Plessner, crucially determines the difference between plants, animals, and human beings. Plessner borrowed from biologist Hans Adolf Eduard Driesch the concepts of open form and closed form to explain the dissimilarities between the vegetal and animal kingdoms in terms of spatial organization and independence from the environments they inhabit.

According to Plessner, a plant – defined by its open form – cannot yet be recognized as being in a positional relationship with its boundary.

The inside of an open form is not endowed with a center of experience. In other words, a plant is limited by a bodily surface whose behaviors and qualities are not regulated by a consciousness. As a consequence of this structural absence, a plant is a non-independent part of the life-cycle of the environment in which it lives (Plessner, 2006, 244). In more synthetic terms, an open form does not have the ability to structure a conscious relationship with either its surrounding environment or its own organism; its form of being is limited to its surface. In other words, an open form *is* its body.

A basic structuring of a conscious (positional) relationship can start to be identified in, what Driesch called, closed forms. In the case of most animals, defined by that very structure, everything that crosses their bodily boundaries in either direction (eg., in the cases of feeding, mating, defecating, breathing, vomiting, etc.) is mediated by an experiential nucleus (represented in gray in Figure 6.1). According to Plessner, this cognitive center can be identified at the physiological level in the nervous systems of animals, and at the psychic level as the conscious awareness that an animal has of its environment. Plessner observed that, thanks to the mediation and articulation organized by its center, an organism defined by a closed form not only acquires a higher level of coherence in relation to its environment, but, to some extent, also becomes independent from it (Plessner, 2006, 251). Unlike a plant, an animal not only *is* its body, but is also *in* its body (as inner experience) (see Figure 6.1) (Plessner, 2006, 251).

As was epitomized in the experiment on the monkey motor-system outlined at the beginning of this chapter, the use of technical artifacts can mediate the relationship between a center of experience and its

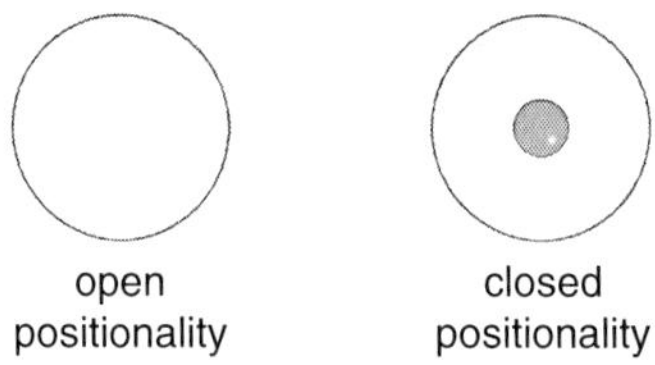

Figure 6.1 The large circles represent the external surface of a body: on the left as the body of an open form that has no autonomous relationship with its environment, as in the case of plants; on the right as a closed form, a type of positionality whose center of experience (symbolized as a grey-filled circle) allows the organism to develop awareness of its world and a degree of independence from it, as is the case with animals who can make decisions with regard to their behavior and in relation to their environment (picture inspired by the De Mul's visual interpretation in De Mul, 2010, 204)

environment. Any form of mediation discloses new possibilities for living creatures for the deepening and the extension of their relationship with their environment.

Plessner's theory of positionality does not, however, only apply to plants and animals. His theory was mainly devised to give a solid conceptual footing to a philosophy that would be able to overcome the shortcomings of a dualistic perspective. According to Plessner, human beings innately establish a relationship with their bodily boundaries. This relationship is more complex than the positionality structured by animals in their closed form. Humans can maintain a central relationship not only with their individual exterior limit, but they are also capable of establishing a cognitive relationship with their very experiential center. Human positionality can, thus, be initially defined as closed: the human animal has a body whose relationships with the world and with other beings are directed by its nervous system.

Plessner observed that being human is, however, also characterized by the capabilities of both self-reflection and self-objectification. These possibilities transcend the simplicity and the directedness of the way that animal life forms establish a relationship with their environment and their body. After a certain age, human beings are intuitively aware of their center of experience. It is this awareness in particular that sets mankind apart from plants and animals. Human positionality features a second experiential nucleus, a second, parallel aspectivity that, to be able to self-reflect on its original center, needs to be external to it (and to a degree separated from it, and thus off-center or eccentric) (see Figure 6.2).

Please note that, in Figures 6.1 and 6.2, the dark grey color represents an experiential center, a cognitive nucleus that is relative to a body. The

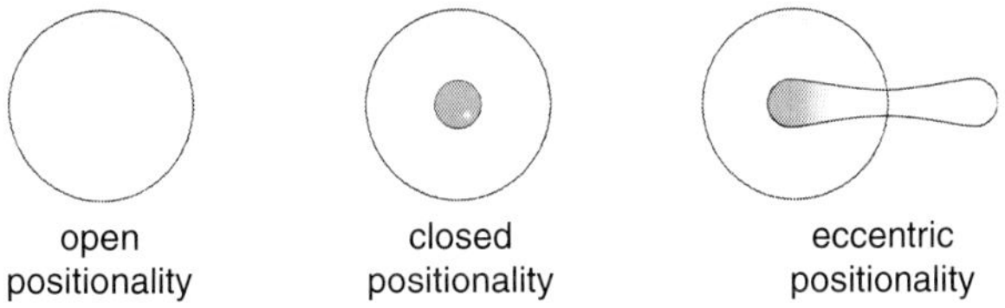

Figure 6.2 The three possible levels of the organic according to the theory of positionality proposed by Plessner in 1928 and inspired in their visual representation by De Mul (De Mul, 2010, 204). The eccentric positionality of human beings, features two coexisting and connected nuclei: one within the body (the bodily experiential center accountable for inner experiences) and the other outside (an aspectivity that allows for the possibility of self-reflection and for interpreting one's own body as an object)

fact that the external aspectivity of the eccentric positionality is not colored indicates precisely that it is not an embodied experiential center, but is an abstract standpoint that does not correspond to a body.

Similar to what was observed in the case of beings defined by a closed form, the mediation of the experiential center in the eccentric positionality grants the human animal the possibility of enhancing and extending one's own body-schema beyond his or her natural effectors.[4] As a common experience in everyday existence, humans can develop a close familiarity with tools (from the proverbial hammer, to the layout of a keyboard, to the dimensions of a car) to a point where, with practice, these technical artifacts are perceived and used as native effectors. This characteristic aspect of being-in-the-world is what Heidegger labels as encountering objects that are ready-to-hand (*Zuhanden*) (Heidegger, 1962, 137–143 / SZ, 104–110).

According to Plessner's theory, beings that are defined by an eccentric positionality are capable of establishing separate and coexisting relationships with both sides of their constitutive boundary, the interior and the exterior (Plessner, 2006, 126–132). In other words, humans are innately defined by a structural ambiguity in the way they relate to themselves: "as eccentric beings we are not where we experience, and we don't experience where we are" (De Mul, 2010, 196).

From this brief outline of the key concepts of Plessner's anthropology, it appears that the particular positionality that characterizes a human being is basically determined by its innate duality, which Plessner terms duplicity. Despite proposing a perspective that is still structurally dualistic, Plessner openly rejected the Cartesian extreme epistemological dichotomy between *res extensa* and *res cogitans*. Instead, he understood the dimensions of interiority and exteriority as distinct aspects of what is really a psycho-physically unitary organism, a living thing.

According to the theory of positionality, a human being *is* a body, is *in* a body, and is, at the same time, outside his or her own body. Specifically, in Plessner's work the eccentric cognitive center that characterizes human positionality is metaphorically located "behind oneself" (Plessner, 2006, 312–317). Similarly, Heidegger also indicated something along the lines of a double temporal positionality as the essential structure of how human beings are in the world. While Plessner imagined the second cognitive center of a person to be spatially behind oneself, according to Heidegger, this second aspectivity is always presented as temporally "ahead-of-oneself."[5]

Each on their own terms, both Heidegger and Plessner structured philosophical perspectives that recognized in the inherent duplicity and ambiguity of the human cognitive structure the fundamental cause of the groundlessness and incompleteness that characterize the human condition. It is precisely because of this particular complication – the insurmountable chasm within oneself – that human beings experience themselves simultaneously as objects and subjects. The characteristic of self-reflection and the possibility of objectifying oneself mean that human existence is not univocally established by nature (as is the case for closed forms) and carried out instinctually. Rather, human existence has to be constantly developed, reshaped, and redirected. The human condition is, thus, always an open project, an open question (Plessner, 1980–1984, IV, 383–385). In this sense, the whole of Western civilization could be interpreted as the interconnected set of philosophical, religious, artistic, social, and productive systems that compensate for the finitude and the uncertainty (as well as the consequent dissatisfaction and anxiety) that are constitutive to being human. In Plessner's work, the artificial dimension of culture is presented as second nature for human beings, as the general context where human beings "make something of themselves" and pursue the balance and completeness to which they inherently aspire (Boccignone, 2009, 5). In this second nature mankind is supposed to find its rootedness, a stable ground that is not possible to attain in the context of its first, fractured, nature.

In line with the general interpretation of culture outlined above, Plessner's *The Levels of the Organic and Man* presents technology as a constitutive part of culture that is inextricably linked to the eccentric positionality of human beings. As such, Plessner understands technology as an ontic necessity of mankind (Plessner, 2006, 344).

> Man tries to escape the unbearable eccentricity of his being, he wants to compensate for the lack that constitutes his life form. Eccentricity and the need for complements are one and the same.... In this fundamental need or nakedness can be found the *movens* for everything that is specifically human, the focus on the *irrealis* and the use of artificial means, the ultimate foundation of the *technical artifact* and that which it serves: *culture*. (Plessner, 2006, 334, my English translation from the Italian edition)

On these premises, Plessner elaborates his understanding of the human being as the "apostate of nature" or, as explained in his first anthropological law, a creature that is "artificial by nature." What Plessner means is that the need for complementation of an unnatural kind is structural to who we are as a species (Plessner, 1980–1984, IV, 382–385). He believed, as is contextualized in the quote above, that it is the destiny of mankind's positional structure to aspire to the transcendence of its original (natural) condition through the establishment of new, artificial worlds (Plessner, 1980–1984, IV, 385).

The interpretation of any form of technology as an artificial way to compensate for human incompleteness and finitude is not unique to Plessner's thought. This idea is relatively common in Western thought and, in particular, in fields such as media philosophy and the philosophy of technology. This interpretation was subject to a radicalization and a sudden increase in its academic significance given the proliferation and progressive integration of digital technology into social processes and practices. Instances of this frame of reference can often be encountered in this text. The transhumanist movements prophesize, for example, a future where culture understands and promotes technology as the evolutionary complement and continuation of biology on the foundations of the Platonic and Cartesian dreams of immortality and detachment from the material world. From a similar philosophical standpoint, in his 1993 book *The Metaphysics of Virtual Reality*, Michael Heim defines the mental dissociation that is entailed in interacting with virtual technologies as "Platonism as a working product" (Heim, 1993, 88).

6.5 The digital poly(ec)centric positionality

As anticipated in the previous section, this book considers that Plessner's heritage and insights constitute a potentially more efficient, balanced, and original foundation for understanding the effects of digital simulation than those offered by the dualistic tradition or by traditional phenomenological perspectives. However, a critical aspect of Plessner's philosophical anthropology jeopardizes his promising contribution to contemporary philosophy of technology, media philosophy, and game studies. This problematic aspect consists of the exclusion of the possibility for historical developments of life beyond the eccentric positionality from his theoretical work. In other words, Plessner considered technology to be an anthropological factor capable only of having a cognitive and psychological influence on the human condition, thus essentially ignoring its evolutionary effects. According to Plessner, the eccentric positionality is the ultimate stage of development of the organic, a stage beyond which it is impossible to attain further advancement. An overcoming of the eccentric positionality, Plessner wrote, "is impossible, as the living thing is now really positioned 'behind itself'" (Plessner, 2006, 315).

With the objective of re-thematizing and supplementing Plessner's positionality theory in the age of virtual and tele-technologies, De Mul proposed the addition of an additional positional form, the extra form of intentional boundary-realization. De Mul explained that this new positionality emerges from the structural hybridization between human biology and interactive digital technologies capable of displacing human cognition, such as telepresence and virtual reality (De Mul, 2010, 193, 194). De Mul's addition to Plessner's work was motivated by the recognition of the crucial anthropological importance of the possibility provided by virtual and tele-technologies to partially outsource the human center of experience (De Mul, 2010). Looking through the lens of Plessner's philosophical anthropology, De Mul understood the cognitive effects of the digital technological mediation on human positionality as those of a technological objectification of its eccentricity (De Mul, 2010, 202). More specifically, De Mul recognized that virtual and tele-technologies, rather than simply displacing the human intentional boundary-realization, were capable of multiplying it. For this reason, De Mul named this additional form of intentional boundary-realization "poly(ec)centric positionality" (De Mul, 2010, 202). He further explained that, on a psychological level, the mediated multiplication of one's center of experience should be understood as a dissociation.

As observed in Plessner's first anthropological law, the construction of the individual human existence is characteristically influenced by the technologies and artifacts that mediate one's relationship with reality. This foundational anthropological tendency is also apparent

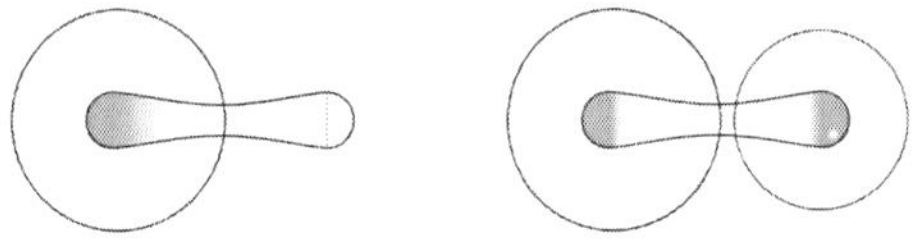

Figure 6.3 Whereas in eccentric positionality only one of the two centers corresponds to a body, in the case of poly(ec)centric positionality both centers are colored, indicating that bodily cognition is characteristic of both. In the poly(ec)centric positionality, the second pole of the eccentricity can also, accurately, be understood as a complete and additional, artificial experiential center. It must be further noted that, in the case of the poly(ec)centric positionality, both experiential centers, both poles of the positional structure are colored. In the eccentric positionality, instead, all the possibility for conscious experience (represented by the grey colour) is concentrated in the one experiential center endowed with a body (picture inspired by De Mul's visual interpretation in De Mul, 2010, 204)

in the relationship that mankind establishes with the digital platform. Especially on the web and in the virtual worlds of videogames, the sense and the structure of identity are malleable and fleeting; a particularly conspicuous manifestation of the fundamentally open and permanently under construction qualities of human existence. Through computer mediation, humans are, in fact, capable of experiencing what can be understood as both an extension and a fragmentation of their agency and identity. It should therefore not be surprising, observed De Mul, that the growth in the number of dissociative mental illnesses has a direct correlation with the increased social diffusion of interactive digital media. In a private conversation, De Mul further explained that, in his perspective, the dissociation that characterizes the poly(ec)centric positionality does not favor a reconciliation of the internal fragmentation inherent to human existence. On the contrary, he added, humans multiply their removal from themselves.

I would like to focus the attention of the reader briefly on a qualitative difference between the mediated augmentations and fragmentations that take place through a poly(ec)centric cognitive structure and the extensions that were observed in relation to the incorporation of artificial effectors in the body-schema of both closed and simple eccentric positionalities. In the case of the incorporation of artificial effectors in the body-schema in closed and eccentric positionalities, the functional purposes that instigated the changes were inevitably actual ones; they were motivated by needs and applications that were relative to the actual world. From this perspective, traditional forms of technological mediation cannot be separated from the actual laws of physics, social conventions, the epoch of human history when they were devised, and so on. Instead, in the case of poly(ec)centric positionality the worlds and the perceptual, cognitive, critical, and operational possibilities disclosed and mediated by virtual and tele-technologies no longer necessarily depend on any notion of actuality.

If poly(ec)centric positionality can be interpreted as a dissociation – as a divergence from the human kinds of ontologies that were developed and established in relation to the actual world – on a psychological level, physically it can be understood as effectively disclosing experiential access to artificial, supplementary bodies that exist simultaneously with the biological body. The reason for the qualifier supplementary when talking about virtual, embodied extensions of the self resides in the fact that the bodies afforded by virtual technology cannot (or at least cannot yet) fully override and replace the actual ones. A similar observation, based on quantitative social research, was presented in Nick Yee's 2014 book *The Proteus Paradox*. Yee offers experimental evidence that psychological biases and dispositions originally developed in relation to the actual world are carried over to virtual worlds in terms of behaviors and superstitions. "To cope with the inundation of information, our brains have developed automated heuristics," wrote Yee, and thus "when we encounter new media and technological devices, we fall back on the existing rules and norms we know.... And when we enter virtual worlds, this mental baggage hitches a ride with us. We react to digital bodies the way we react to physical bodies" (Yee, 2014, 54).

In the context of adopting Plessner's theory of positionality to understand how the experience of virtual worlds is influencing human kinds of ontologies, it must be noted that Plessner's work offers the initial advantage of presenting cognitive processes as having a necessary dependence on an (actual, biological) bodily substrate. With a specific focus on the ontological effects of the experience of virtual worlds, this book has argued that digital simulations need to be recognized as uniquely extending, distorting, and fragmenting the perceptual, cognitive, critical, and operational capabilities of human beings. This augmentation allows us to experientially transcend what is actually present and can contribute to shaping our thoughts and behaviors in virtual contexts that are no longer exclusive and inflexible, but are themselves elements that can be designed and manipulated. This shift toward modality and projectuality grants human beings the possibility of structuring virtual ontologies that are (at least conceptually) independent from their possibilities of being employed to understand or categorize the actual world, or from their efficiency in determining who we are and how we behave in the world we share as biological organisms. In this way, as anticipated in the opening chapters of this book, virtual worlds can be recognized as the contexts in which the ways we construct ontologies can transcend the actual world, and where a new, broader humanism has already begun to arise.

To put this understanding in historical perspective, at the beginning of the last century, futurist writer Filippo Tommaso Marinetti predicted that the "futurist transhuman" would be "multiplied by the machine." This new being was to be a hybrid creature, "with replaceable parts" and "multiple and simultaneous awarenesses" (De Mul, 2010, 32). Similarly, Sherry Turkle more recently wrote that virtual worlds "encourage us to think of ourselves as fluid, emergent, decentralized, multiplicitous, flexible, and ever in progress" (Turkle, 2007, 263–264).

As illustrated in Figure 6.3, the cyborg that emerges from the hybridization of biology and digital technology is characterized by a positionality that departs from a simple eccentric structure. The reaching of the new, poly(ec)centric positionality has, I argue, a series of far-reaching implications. The most fundamental aspect of this novel anthropological stage is that the external pole of its eccentricity is also defined by a boundary, by a body. In virtual worlds or through the mediation of telepresence the affordances and possibilities offered by the digital extensions of the cybernauts' bodies become, with use, effective parts of their actual body-schemas. This merging – or rather this multiplication – is not only the basis for several social practices that already involve acting in virtual worlds and experiencing virtual phenomenologies as a constitutive component (e.g., professional videogaming, or the training of aviation pilots and surgeons), but is also a common experience in the everyday interaction with digital media, where virtual interfaces (HUD, GUI, videogame avatars, etc.) and hardware controllers become familiar extensions of our embodied interactive structures. In the context of interactive digital mediation, the concept of presence can thus be understood as a consistent feeling of immersion in a simulated world, the intensity of which depends on the smoothness and coherence of the incorporation of one's virtual body-schema into one's actual (or, better, biological) body-schema.

In sum, from an anthropological perspective on media philosophy, the relationships that human beings establish with their bodies, when their bodies are complemented and extended in virtual worlds, must be embraced as defined by multiple, coexisting experiential centers corresponding to their artificially multiplied bodies, regardless of their organic, telepresent, or virtual constitution. In accordance with what was observed, and in analogy with the qualities that were recognized as defining a world as in the first chapter (in a nutshell, the need to be persistently perceivable and behaviorally consistent for it to be intelligible), we can conclude that, in order to elicit ontological effects, a simulation needs to be possible to be, at least to some degree, incorporated by its users/players.

6.6 Political cyborgs?

In the incomplete philosophical project of which *Being and Time* was an initial part, Heidegger ventured to answer what he considered to be the single most fundamental and overlooked question in Western philosophy since its inception in ancient Greece: "What is the meaning of Being?" Heidegger explained, in the introduction to *Being and Time*, that the question of being can only be addressed once the being for whom the question of being is important is thoroughly analyzed and understood. *Dasein* (literally, being-there) is, for Heidegger, the being asking the question of being – the being for whom being matters. Consequently, to provide a solid footing for the decisive question he was trying to answer, he started his masterwork by analyzing the fundamental characteristics of *Dasein*. With these objectives in mind, the first section of *Being and Time* was structured around the most essential dimension of *Dasein*, its temporality (Heidegger, 1962, 32–33 / SZ, 12). Heidegger's analysis of the human being can be understood as an essential understanding of human existence and its finitude. In the last century, the level of abstraction and the focus on the individual temporality in Heidegger's philosophical work were critical points in the reception and diffusion of his work. These points have also led to the long-standing reputation of Heidegger's thought as a highly speculative philosophy that is difficult to apply to actual socio-political issues and questions.

As a case in point, the public dimension of being human plays only a very marginal role in Heidegger's *Being and Time*. In the early developments of his thought being in society is reductively understood as a moment when a human being's authentic, individual self is contrasted with a social mask that he or she is forced to wear, leading to an inauthentic mode of existence. Instead of adhering to personal inclinations and aspirations (and thus exerting authentic will), in the public sphere, one acts impersonally. In public, regardless of the specific social context, the individual is conditioned to behave in ways and according to norms that are expected of him or her in that context, "that is what one does." Building on these perspectives, which are heavily influenced by Heidegger's Lutheran upbringing and are further motivated by Heidegger's disappointment with the Weimar republic fiasco, *Being and Time* generally belittles the social role and the relevance of political debate, characterizing it as unconstructive chattering (Heidegger, 1962). Heidegger's disregard for the social dimension and for political involvement can be interpreted as the secularization of the ideological separation operated by Martin Luther between the public and private spheres – the private sphere being where salvation could be earned according to this iconic figure of the Protestant Reformation.

Similarly to Heidegger, the early Plessner also explicitly distanced himself from any radical notion of community, distinguishing the sphere of intimacy from the public façade that was necessary for individuals to meet on the common ground of social relationships (Boccignone, 2009, 2, 3). In his mature writings, Plessner recognized a necessary, indispensable role for both dimensions of the human *Doppelgänger* (with reference to the human being's dual, eccentric positionality). In strident contrast to the early Heidegger's deprecation of both public life and an involvement in political life, the later Plessner no longer focused on a single aspect of the social/individual dichotomy. For the later Plessner, it was unreasonable to consider only one aspect as worthy of philosophical attention while marginalizing the other as idle and inauthentic. He further condemned the intellectual elites' abandonment of the communal sphere as being among the crucial factors that allowed racist biology to become the dominant ideology of an authoritarian state (Plessner, 1959). In Plessner's later theoretical framework, the socio-political sphere is understood as one of the possible contexts in which man, the "apostate of nature," can collaborate and take responsibility in setting up a new world, giving it structure and promoting sociocultural change on a larger scale than the individual (Plessner, 1980–1985, IV, 395). For Plessner, the role of politics is precisely to engage humans in the endless and inscrutable process of structuring their artificial "homeland."

Of the wide variety of topics treated in Heidegger's work, one of the most directly involved with social practices (or social policies) is his dread of the danger of human objectification that he saw as ensuing from the growing sway of the technological mindset. In contrast to Heidegger, Plessner, in his work, opposed the idea of a direct correlation between the diffusion of technologies in social practices and the risks of alienation or segregation. Arguing against the validity of a causal connection between artificiality and alienation, Plessner claimed that the alienation theorem is mistakenly founded on the Romantic understanding of the relationship between individuals and society, according to which people can only find their (natural) balance when harmoniously integrated into their respective communities (Boccignone, 2009, 3). From Plessner's perspective, it would be illogical to blame the mechanization of production or the social diffusion of technology for human alienation. The isolation and the dissatisfaction that characterize the human condition are better explained, according to Plessner, as derivations of the fundamentally broken constitution of the human being, rather than being caused by the specific qualities of technological environments. As explained in Plessner's first anthropological law, our cultural and technical environments are, to some degree, one and the same with segregation and alienation; they are all by-products of the characteristically incomplete and project-oriented positional structuring of human beings.

A similar understanding was offered in the first chapter of this book, where technology was described as "a conglomerate of technological artifacts, specific forms of knowledge and capabilities ..., the necessary geographical and social infrastructure, economic interests and societal norms and values" (De Mul, 2002, 30). Because the proposed definition of technology is heavily tied up with its involvement with sociocultural processes and its methods of use, it would be inconsistent to treat technology as machinery or infrastructure that can be understood and thought of apart from its contexts. It is evident that the questions concerning digital technology cannot be thoroughly understood, or even properly formulated, without attending to the relationship between technology and social practices. Clearly, a direct and acritical application of Heidegger's early frameworks in the philosophy of technology would be unable to incorporate this socio-political dimension of (digital) technology adequately. Accordingly, I propose an integration of Heidegger's original insights with a specific treatment of the possible roles of the digital outsourcing of one's consciousness in steering or facilitating social change.

The brief, initial treatment of the socio-political dimension presented in this book is not intended to be exhaustive or definitive. The topic is too vast (and too interesting) to treat in full in the context of this book, where it is merely a contributory topic to a more speculative philosophical project. However, for the sake of completeness, I will introduce a general outline of the possible horizons that the cultural shift toward projectuality in virtual worlds discussed so far could entail in the context of socio-political processes. In providing such an outline, I will refrain from taking a large-scale perspective on the relationships between the social-productive use of technology and political power. Notwithstanding the functional merits of a systemic approach of that kind, I believe that the most appropriate angle from which to analyze the social and political effects of digital mediation would be that of focusing on the micro, anthropological level. From my theoretical standpoint, the political consequences of augmented ontologies are best analyzed on an individual scale and observed as the modification of the relationship between people's private and public lives.

In my understanding, the possibility of fragmenting, extending, and distorting traditional human kinds of ontologies by accessing and experiencing an assortment of virtual alternatives to the way worlds are customarily perceived and organized can take two directions in terms of individual participation in the socio-political debate:

1. the fluidification of the thought process as a consequence of its polarization toward projectuality, and

2. the trivialization of an involvement with the actual world stimulated by the combinatorial and malleable qualities of digital ontologies.

These two envisaged effects of virtual technologies on an individual's engagement in the socio-political aspect of his or her existence should not be understood as alternatives to one another, but rather as coexisting and mutually influential. Interestingly, a definite parallel can be traced between the two proposed understandings of the political role of digital media and the two myths that were recognized as underlying the modernistic understanding of the social relevance of art.[6] The following two sections offer a more detailed explanation of both effects.

6.6.1 "Fluidification"

The ideas leading to this first understanding of the potential sociocultural effect of the diffusion of digital technology in social practices originally emerged in debates concerning aesthetics and theories of art during modernism. As was notably contextualized in Walter Benjamin's 1936 essay, "The Work of Art in the Age of Mechanical Reproduction," the artistic production of Dadaism, in particular, challenged traditional representational norms and the classical social and political roles of art. The aesthetic rhetoric of Dadaism intended to demonstrate that no object or set of values could legitimately be privileged or considered ontologically superior to any other. The general insubordination of modernistic currents to artistic conventions and, metonymically, to the established univocality and stability of both Western culture and traditional worldviews, were overtly avowed in various modernist manifestos. In this sense, Dadaism embraced artistic expressive forms as catalysts of social change.

It is not a coincidence that pre-digital works of recombination were first introduced in the history of art by Dadaism. Traces of their French-dada origin are evident in the names of practices such as assemblage, collage, and photomontage. A remarkable contribution to the Dadaist movement, especially in terms of its politically subversive stance, came from German photomontage artist John Heartfield.[7] Heartfield's politically insubordinate works comprised reassembled black-and-white photographs paired with seditious slogans. In his compositions, Heartfield proposed alternative visions of reality through the newly introduced technique of photomontage, allowing him to achieve a form of persuasion that was more disquieting than any previously utilized artistic (or propagandistic) form.

In a way that is conceptually analogous to how the Impressionistic currents (and, earlier, the Dutch Renaissance) demonstrated the possibility of conceiving and producing works of art without the necessary involvement of mythology, religion, or themes that idealized or justified power, Dadaism embraced works of art in ways that were supplementary to (or even dissonant with) aesthetic gratification. Disentangled from both the tastes of a specific social group and dependencies on structures of power (both religious and political, as Benjamin noted in relation to the sociocultural role of traditional artistic expression in "The Work of Art in the Age of Mechanical Reproduction"), artistic production could take an independent ethical stance and even an openly propagandistic dimension.

In the twentieth century, similar efforts to the ones listed above aimed at making social processes more malleable and subject to fluid changes (from which were derived both the name of the artistic current Fluxus and the title of this section) became the preferential strategies through which artists and philosophers developed and elicited their sense of possibility and their subjunctive mood: their "ability to conceive everything that there might be just as well as to attach no more importance to what is than to what is not" (Musil, 1996, 11). Alexander R Galloway has argued that, not unlike how Heartfield harnessed the persuasive power of traditional media, "video games do nothing but present contemporary political realities" and that, differently from traditional and passively experienced forms of mediation, they "achieve a unique type of political transparency" (Galloway, 2006, 92). Analogously, Bogost has argued that virtual worlds, thanks to their specific behavior-based mode of representation and interaction, offer a new and peculiarly momentous form of engagement in socio-political discourses. Bogost labelled the unique and novel form of persuasion that can be achieved by means of interactive digital media "procedural rhetoric." As the name itself implies, the persuasive strategies and the rhetorical potential of procedural rhetoric are directly dependent on the capability of the digital medium to run processes and execute rule-based symbolic manipulation (Bogost, 2007). Two qualities of rhetoric as disclosed by simulational media are particularly interesting for framing the fluidifying effects of the experience of virtual worlds from the humanistic angle adopted in this chapter. The first of these is procedural rhetoric's general orientation toward modality, and the second is its characteristic interactivity.

Using Arthur Schopenhauer's notion of transcendental will to observe and frame the interaction between the player and Link (the protagonist of the Nintendo videogame series *Legend of Zelda*), philosopher Dario

Compagno has explained that a crucial difference between expressing our will in the actual world, as opposed to expressing it in virtual worlds, resides in the concept of choice. According to Compagno, people do not really have freedom of choice in the world indexed as actual simply because, in the actual world, they cannot know in advance the exact outcome of their possible choices. In the temporally linear perception of our existence, Compagno argued, the options for action available to us, although possibly based on intuition or on the rationalization of our past experiences in analogous situations, are always blind to their consequences. In his interpretation, these are not real choices, but hopeful decisions that have the quality of being irrevocable. In single-player videogames, instead, each alternative possibility for action, regardless of its narrative or performance-related relevance, can be explored in minute detail through gameplay, or rather through the possibilities for reverting causality and for restoring the game state that are granted by the formal, modal nature of digital processes.

Since the release of the first interactive digital entertainment titles, videogames have allowed for basic ways to reset or revert the casual-temporal state of their worlds. These possibilities were available to the players either in the form of save game options or through the ability to reset and restart the virtual experience. With simulations and games becoming more sophisticated in the 1980s, both in their game design and from the point of view of their hardware substrate (e.g., with the larger amount of memory available on personal computers and home consoles), the videogame industry witnessed the emergence of new ontological affordances like those offered by checkpoints. Checkpoints are particular moments in the time–space progression of the experience of linear (or semi-linear) videogames with a function in gameplay similar to the function of bookmarks to the activity of reading. Checkpoints grant the players the possibility of resuming their activity from a specific point in their advancement or of exploring and replaying a specific section of content they had previously encountered without having to restart the experience from the beginning.

More modern game releases, like *Legend of Zelda: Majora's Mask* (Nintendo EAD, 2000), *Blinx: The Time Sweeper* (Artoon, 2002), *Prince of Persia: Sands of Time* (Ubisoft Montreal, 2003), and *Braid* (Number None, Inc., 2008), embraced time-reversal, as well as the possibility of playing the same game section over and over until satisfied, as a fundamental feature of their gameplay. The exemplary user-controlled, smooth manipulation of game causality that these titles have in common (to different extents and for different scopes) reveal to the players the virtual consequences of any of their actions and, at the same time, grant them the chance to visualize, manipulate, and choose their desired course of action for each event in the game. In other words, for the first time in the history of culture, through videogames and simulation featuring interactive time-manipulation mechanics, human beings can truly choose what to will.

Compagno's perspective encapsulates two essential components of rhetoric as experienced through virtual technologies. First, the contents of digital simulations inevitably show – in their several different ergodic possibilities and ramifications – that the current state of things could be different from what it actually, presently is, effectively fluidifying human thought. My argument is that, when visualizing several alternative possible states of things, and when manipulating and becoming familiar with these states, the user of interactive, digital simulations is likely to become more prone to embracing what could be over what is. In other words, I believe that the diffusion of digital mediation and a simulational mindset in social processes favors the polarization of culture toward modality. The cognitive flexibility that could emerge from being in virtual worlds could, thus, facilitate both the envisaging of new possible courses of action and their acceptance. In this sense, experiences of virtual worlds are inherently expected to foster the involvement in processes that guide and promote sociocultural change. From this understanding of the social significance of digital mediation follows the idea that any digitally-mediated simulation should be recognized as having some intrinsic political influence. This is a statement that is valid for any digitally-mediated simulation, including those without overt didactical aims, those without a definite social agenda, and even those that do not purposefully materialize specific ideologies. Virtual worlds – all virtual worlds – cannot avoid rehearsing the worldviews that shaped their technological structures, as well as the beliefs and values deriving from the cultural context in which they were designed.

The second aspect that I believe is foundational for a thorough, humanistic understanding of the fluidifying effects of the experience of virtual worlds on human kinds of ontologies is the very interactivity of virtual worlds. As a defining quality of a virtual experience, interactivity allows for both the exploration of virtual possibilities in digitally simulated worlds and the in-game expression of the player's will, as summarized above. Similar to the position toward the mediation of thought held by both Socrates and Wittgenstein, I believe that interactivity allows for a less abstract and univocal kind of rhetoric and a more dialectic and flexible approach to ideas than text or passively experienced forms of mediation could ever structurally offer. Notably, as was the case for the oral elaboration and transmission of thoughts and values, the interactivity and procedurality of computers afford a dynamic negotiation of meaning that cannot be attained with pre-digital forms of mediation.

6.6.2 "Trivialization"

Gonzalo Frasca has noted that a belittlement of the (actual) historical process could ensue from an interpretation of existence as analogous to a simulation, as a branching system of possible alternatives:

> A video game about Anne Frank…would be perceived as immoral, since the fact that she could survive or die depending on the player's performance would trivialize the value of human life. We all know that Anne Frank died and the reason for her death; her story serves to convey a particular set of values. (Frasca, Gonzalo, in Wardrip-Fruin, Harrigan, 2004, 86)

The trivialization of values that Frasca discussed is not unique to videogames. Rather, it is an effect that is common to all forms of experience characterized by combinatorial or modal possibilities, such as engaging ergodic literature, playing card games, or exploring a digitally simulated world. In proposing a combinatorial, modal interpretation of reality, these experiences intrinsically suggest the latency and the validity of states of the world that are potential and alternative to the one indexed as actual.[8]

The belittlement of the actual historical process and of culture and life itself (I believe that the vilification of life in particular, both one's own and that of others, is especially common in videogames) by means of (countably) infinite instances of possible permutations was also the central theme of Borges' 1941 short story, "The Library of Babel," which describes an unthinkably vast library consisting of seemingly endless desolate, interconnected hexagonal storage shafts. Together, these storage shafts contained every possible 410-page volume that could be composed by reconfiguring the letters of the alphabet and certain punctuation characters (Borges, 1994). In an earlier (1939) essay, "The Total Library," Borges anticipated the themes that would later constitute "The Library of Babel," in particular the idea that a boundless, combinatorial library would contain every imaginable combination of words and letters bound into books, regardless of the intelligibility of these combinations. Such a combinatorial library would contain

> [e]verything: the detailed history of the future, Aeschylus' *The Egyptians*, the exact number of times that the waters of the Ganges have reflected the flight of a falcon, the secret and true name of Rome, the encyclopaedia Novalis would have constructed, my dreams and half-dreams at dawn on August 14, 1934, the proof of Pierre Fermat's theorem, the unwritten chapters of *Edwin Drood*, those same chapters translated into the language spoken by the Garamantes, the paradoxes Berkeley invented concerning Time but didn't publish, Urizen's books of iron, the premature epiphanies of Stephen Dedalus, which would be meaningless before a cycle of a thousand years, the Gnostic Gospel of Basilides, the song the sirens sang, the complete catalog of the Library, the proof of the inaccuracy of that catalog. Everything: but for every sensible line or accurate fact there would be millions of meaningless cacophonies, verbal farragoes, and babblings. Everything: but all the generations of mankind could pass before the dizzying shelves – shelves that obliterate the day and on which chaos lies – ever reward them with a tolerable page.... I have tried to rescue from oblivion a subaltern horror: the vast, contradictory library, whose vertical wilderness of books runs the incessant risk of changing into others that affirm, deny, and confuse everything like a delirious God. (Borges, 2001, 216)

When interacting with virtual worlds, people can experience the same confusion as the imaginary visitor in "The Total Library." In terms of their ontologies, being exposed to a multitude of incoherent and often bizarrely unworldly virtual experiences, can contradict, confuse, and trivialize traditional values, ideas, and beliefs while asserting the triviality of any form of knowledge. Accordingly, Nicholas Carr upheld in his 2010 book, *The Shallows*, that the proliferation of media channels requesting the attention of the users and prompting them for interaction in today's increasingly digitally-mediated existence necessarily entails that attention and concentration are thinly spread. For Carr, this sociotechnically-enforced lifestyle is making us less focused and less prone to engage in deep reasoning. By analogy, it is also imaginable that the distribution of one's self (in terms of time and care) among an assortment of virtual worlds will lead to a diminished participation in actual socio-political spheres. The existence of foundations aimed at saving certain virtual creatures (the Morlocs in Blizzard Entertainment's *World of Warcraft*) from extinction and many examples studied by Yee (2014) in relation to the absorption and dedication with which online players work to organize guilds and build online communities are suggestive of this tendency.

Following from these insights and interpretations, I argue that the partial dissociation from the world that humans biologically depend upon (and with which they structured their original ontologies) must also be recognized as having a compelling anthropological appeal. This appeal is particularly evident and especially momentous in the case of videogames, and I believe this is one of the key reasons why game-like simulations are currently employed in a wide range of rhetorical contexts, from advertisements to political propaganda. I see this appeal as emerging from the fact that the logical and ontological structures of videogames are often explicitly designed to elicit feelings of self-realization in players. Such feelings, as recent trends in casual game development have demonstrated, can be triggered in a multitude of ways, including stimulating actual development in the performance skills of the player, conveying a sense of progress through the constant growth of player avatars or through the development of in-game narratives, or using extrinsic motivators and psychological reinforcers (Mosca, 2012).

The focus in the design of virtual experiences outlined above resonates with the recognition of self-construction and self-realization as basic ontological needs for beings, such as humans, who are defined by the eccentric positional structure proposed by Plessner. This interpretation of the social role of (ludic) virtual technology is also reminiscent of both Lévy's understanding of virtuality and of what Plessner identified as the "focus on the *irrealis*" as a constitutive aspect of being human (Lévy, 1998; Plessner, 2006). In other words, humans can be identified as being almost invariably attracted to certain self-fashioning activities, as well as to games, (a drive that can often turn into a compulsion) precisely because those practices offer rewarding answers to the unanswerable questions at the core of humans' inherently broken existence. They deliver opportunities for people to projectually re-construct and aesthetically refashion themselves in ways that are significantly more accessible and less inertial than their actual, embodied counterparts.

It is relevant to observe, here, that the thematic and marketing decisions behind commercial videogames can easily be demystified as promising the fulfillment of precisely those fantasies and drives that are hard to attain, illegal, or physically impossible in the world that human beings share as biological organisms. In this way, commercial videogames generally offer clear and openly quantifiable objectives that promote feelings of growth and meaningful progress.

6.7 Concluding remarks

Pivoting on the need for completeness, balance, and grounding that were identified by Plessner as fundamental ontological needs of the characteristically broken and meaningless human condition, interactive, virtual worlds can be observed and utilized as ontological instruments. The phenomenological and epistemological shift afforded by interactive, virtual technologies cannot take place without ontological consequences; consequences that, as Flusser observed, are ushering humanity away from a state of subjectivity into one of projectivity. The ontological effects and changes fostered by the diffusion of virtual worlds in social practices entail, in turn, socio-political consequences. The ontological effects of the introduction of interactive, digital technology in socio-political processes mostly proceed from the changes they elicit in individuals. From this micro, anthropological standpoint, virtual technologies were recognized as possessing a fundamental democratic power, a power that is fundamentally ambiguous and, from a certain point-of-view, even contradictory. The interactive experiences of virtual worlds, together with their characteristically combinatorial and procedural processes, can in fact be seen as both

- facilitating and encouraging individual engagement in the socio-political sphere, and
- denying and confusing the ontological superiority of the world indexed as actual over a myriad of virtual ones. This levelling in value comes with a momentous belittlement of the historical process and of existence itself.

Understood from the proposed perspective, all virtual worlds can be deemed as holding an implicit political relevance that is a derivation of their combinatorial, modular, and self-organizing constitution. Both the use and the design of virtual worlds as means of production are, thus, implicitly political activities.

7
Virtual Worlds as Poetic Allegories

7.1 Introduction

According to sociologist Neil Postman, all media forms generally function

> like metaphors, working by unobtrusive but powerful implication to enforce their special definition of reality. Whether we experience the world through the lens of speech or the printed word or the television camera, our media-metaphors classify the world for us, sequence it, frame it, enlarge it, reduce it, colour it, argue a case for what the world is like. (Postman, 2005, 10)

In line with Postman's gnomic observation, my philosophical aim of overcoming the univocal mediation of philosophical ideas as linear, textual information considers digital mediation to be one of the viable contexts in which new rhetorical and dialectical ways of shaping culture and thought can be pursued. The use of virtual worlds as mediators of human thought and human experience (as proposed in this book) can, at least in certain aspects, be associated with the emergence of what Walter J Ong labeled in the eighties (in relation to pre-digital media of mass communication) a "second orality". The term second orality indicates, for Ong "a more deliberate and self-conscious orality, based permanently on the use of writing and print" that is emerging in contemporary culture as a consequence of the development and diffusion of new media forms (Ong, 1982, 136). Similarities between the perspectives on the rise of a second orality and the emergence of augmented ontologies observed in this book can be identified in the following three aspects:

- Digitally mediated content is seen as additive and aggregative in both theoretical frameworks, meaning that it does not develop following institutionally established conventions, but changes and develops following the tastes, preferred uses, and – increasingly – direct contributions from its user communities.
- Both perspectives present thoughts and experiences that are mediated by computers as situational rather than abstract. In particular, these thoughts and experiences are always conceptualized and presented with close reference to how people structure their relationships with the actual world.
- Following both perspectives, the emergence of a second orality does not exclude literacy or aspire to a wholesale substitution of text. The understanding of a return to a dialectic way of establishing meaning in culture is post-literal, in the sense that it is simultaneously different from traditional literacy and inextricably rooted in it.

The possibility for computers to express content through interactive metaphors was introduced in Chapter 3, where digital simulations were presented as intelligible and persistent, designed interactive systems that disclose a source system (or systems) through a less complex, technically mediated system. The relationship between a simulation and its source system (or systems) functions by means of implicit analogy. In line with these ideas, simulations are treated in the present chapter as extended, behavior-based declinations of wider cultural processes, or metaphorism. This chapter also elucidates how the cultural influence of the experience of virtual worlds can be more suitably understood as a poetic, rather than a rhetorical, use of metaphor.

This chapter aims to complete and complement the perspective proposed in this book by providing an anthropological account of what it is like to be human – or rather extended humans – in a fragmented multitude of virtual, ontologically independent worlds. Toward this objective, this chapter focuses on the creative possibilities for humans in their role as creators of meaning via interactive analogies in virtual worlds. More specifically, it explores digitally mediated simulations from two complementary directions:

- First, from the perspective of a media philosophy researcher, I play videogames, analyze the structure of their content in relation to their technological background, and organize claims about their meaning and relevance as factors of sociocultural change.

- Second, as a videogame designer who contributes to the creation of digital worlds, I draw on several examples from videogames that I designed or conceptualized and supervised. The benefit of drawing on these games is that I know their structure firsthand and am able to discuss their design process in detail; this makes them very useful, and hopefully easily comprehensible, didactic examples.

7.2 Rhetoric and poetry

An uneasiness and dissatisfaction with the level of abstraction and other limitations imposed on philosophy by the codification of ideas in written language has long been recorded in the history of thought. The first expression in written philosophy of an awareness of potentially detrimental and constraining cultural and cognitive effects that ensue from the adoption of a written canon for the expression of philosophical thought is found in Plato's dialogue *Phaedrus*, given voice through Socrates' position toward the newly introduced practice of writing. These philosophical concerns, as Heidegger observed, have largely been overlooked by the mainstream currents of Western philosophy.[1] The traditional, and often acritical, identification between thought and its written mediation led, from Heidegger's perspective, to a restriction of the horizon for philosophy and to a "perversion" of how philosophical truth was pursued in the Western world. A similar critical perspective was recognized by Nyíri in Ludwig Wittgenstein's later thought. Based on his interpretation of Wittgenstein's attitude toward written philosophy, Nyíri labelled Wittgenstein a "philosopher of second orality" (Nyíri, 1996; see Section 5.3 – Beyond the exclusivity of language: the philosophical practice of "doing").Other scholars, including David G Stern, have taken a different perspective on Wittgenstein's approach to production and the philosophical use of the textual medium. According to this alternative interpretation, Wittgenstein was not attempting to employ text in a more interactive, dialogic form that would hybridize it with qualities traditionally ascribed to orality. Rather, he was trying to find a novel and more appropriate use for text – a use that would increase its flexibility and allow for a personal access to ideas, anticipating the qualities and possibilities of hypertext.

In the digital era, the criticism of the linear, inflexible, and subjective nature of the textual mediation of thought (and disclosure of experiences) fostered the social adoption of digital mediation as a viable alternative to textuality for the creation, dissemination, and preservation of information. Hypertext, in this sense, is just one among many computer applications that, inevitably inheriting the structural qualities of the digital medium, can overcome certain limitations inherent in the traditional and exclusive textual constitution of Western philosophy. Similar qualities can be discerned in several interactive, adaptive, ergodic, and collaborative uses of computers.

Ian Bogost has proposed a rhetorical understanding of the digital medium, stating that interactive digital entertainment opened up a new domain for persuasion. According to Bogost, this new rhetorical horizon is characterized by its particular expressive and persuasive affordance, procedurality (Bogost, 2007, ix). What Bogost believed is that videogames' procedurality is a language (or rather a form of literacy) constituted of rules (Bogost, 2007, 9). According to Bogost, and other game developers and artists aligning with the proceduralist current, the persuasive power of the language of videogames stems from how the logical, causal, and aesthetic qualities of virtual worlds can discernibly be put in relation to the actual world (see Bogost's 2005 paper "Frame and Metaphor in Political Games").

The rhetorical power of interactive digital mediation can be – and in fact already is – employed in several social processes and practices, including the promotion of commercial products (adver-games), support of political agendas (propaganda games), facilitation of education and training (serious games), and critical reflection on the very ways in which we craft virtual worlds (self-reflexive games). The virtual worlds of videogames and digital simulations can be seen as having the potential to lead to significant long-term changes in society and to strengthen the political awareness and social engagement of individuals. Embraced as metaphors, virtual worlds have the power to disrupt and change fundamental attitudes and beliefs about the world and can thus be used persuasively in the pursuit of institutional as well as artistic, philosophical, critical, and personal goals.

In contrast to the perspectives on digital mediation presented by both Ong and Bogost, I do not propose understanding virtual worlds strictly as a new form of language (in terms of either their oral or literary dimensions). I do, however, acknowledge Bogost's and Ong's point that some structuring logics of content for digitally mediated simulations have affinities, and often a direct correspondence, with several traits of language. For example, both digitally mediated simulations and language are characterized by a necessary degree of intelligibility structured by some kind of semiosis. However, one could problematize the understanding of videogames as forms of textual or literary expression by questioning the proposed identification on a very practical level.

For example, if Nintendo's 1985 videogame, *Super Mario Bros.*, contains a strictly linguistic message, then to what does a level correspond? Is it a sentence? A paragraph? A word? Do different genres express their alleged meaning through what can be understood as radically different languages, or simply through a different syntax? Is the logical structure of a videogame its grammar or its (ergodic) plot? These questions, I believe, cannot find objective answers through a direct conceptual identification of interactive digital worlds with language.

Instead of drawing an analogy between virtual worlds and forms of textual or literary expression, I propose understanding digitally simulated worlds as media; as technologies that can disclose and afford meaning through semiotic processes that are not always identifiable with those of text or spoken language. The subsequent sections of this chapter explain how virtual worlds as media can convey allegorical meanings that do not necessarily entail a linguistic kind of semiosis.

In exploring the computer as an increasingly prominent and socially integrated medium, and assessing the cultural roles and dimensions that characterize digital mediation, I believe it is fruitful to commence our analysis, similar to what Bogost did, from the origins of the concept of rhetoric in ancient Greece. In the second half of the fifth century BC, especially in Athens, the sophists heralded a philosophical perspective according to which humans are imperfect creatures – finite, limited, and thus incapable of pursuing any form of absoluteness. Humans, however, could use language and logic rhetorically to establish consensus in relation to the best perceived course of action for the benefit and/or the advancement of the *polis,* the community of citizens. In Aristotle's treatise, *Rhetoric,* dating to the fourth century BC, the eponymous term is defined as the faculty of observing in any given case the available means of persuasion. In line with this original understanding of the term, scholars have often limited the application of the field of rhetoric to political discourse. Other academics, though, have embraced rhetoric as a general quality of human expression that, potentially, can encompass every aspect of culture.

Unlike rhetoric, the poetic use of language was not originally associated with practical issues or the persuasive endorsement of one political course of action over another. According to philosopher Paul Ricoeur, the fundamental aim of poetry is to "compose an essential representation of human actions; its appropriate method is to speak the truth by means of fiction, fable and tragic *muthos*" (Ricoeur, 2008, 13). In informative, rhetorical, or didactic discourse, units of meaning take the function of signs that either refer to, stand for, or point to something.

In poetry the signs represent nothing, instead they affirm, they assert. Poetry ignores the world that is presented to our senses and evokes its own worlds fictionally. Quoting Sir Philip Sydney, Ricoeur observed that, if it were necessary to compare poetry with something, it would be mathematics: "[t]he poet, like the pure mathematician, depends not on descriptive truth but on conformity to his hypothetical postulates" (Ricoeur, 2008, 226).

Ricoeur's understanding of poetry closely mirrors that offered by Paul Valéry's 1939 essay, "Poésie et Pensée Abstraite" (Poetry and Abstract Thought). For Valéry, poetic language is, in its cognitive function, "the effort which makes live in us that which does not exist" (Valéry, 1939, 1333, translation by Herbert Marcuse in Marcuse, 1991, 68). According to Valéry, poetry breaks the spell of our acceptance of how things are to us as human beings; it is the establishment of a new world. In a very literal sense, this understanding of the ultimate goal of the poetic activity adheres to the original Greek meaning of *ποίησις* (*poiesis*), to create, to give shape.

7.3 Metaphors and allegories

Apart from an initial and succinct explanation of the methodological and cognitive dimensions of poetry, Ricoeur dedicated his *magnum opus* to one poetic linguistic tool in particular, the metaphor. Specifically, Ricoeur focused on the use of metaphors in the structuring of meaning and knowledge. The philosophical treatment of the concept of metaphor was first recorded in the fourth century BC by Aristotle in his *Poetics*, where he defined metaphorism as the process of "giving a thing a name that belongs to something else; the transference being either from genus to species, or from species to genus, or from species to species, or on the ground of analogy" (*Poetics*, 1457, b 6–9). Before going into these structural characteristics, Aristotle highlighted an essential feature of the use of metaphors in language, that of: "setting the scene before our eyes" (*Poetics*, 1410, b 33); "making your hearer see things" (*Poetics*, 1411, a 25, b 10); and "representing things in a state of activity" (*Poetics*, 1410, b 33).

A very similar interpretation of the evocative, poetic power of metaphor was presented by Friedrich Nietzsche in his 1872 book on dramatic theory, *The Birth of Tragedy from the Spirit of Music*. According to Nietzsche, for the poet, the

> metaphor is not a rhetorical trope, but a representative image which really hovers in front of him in the place of an idea. The character is for

> him ... a living person, insistently there before his eyes, which differs from the similar vision of the painter only through its continued further living and acting. (Nietzsche, 2013, 26)

In Ricoeur's analysis, the metaphor serves not as a mere literary ornament, but rather a linguistic device with a cognitive, ontological value in its own right. In his work, Ricoeur clearly defined the dual functions of the metaphor that he identified in its use in culture, distinguishing its rhetorical employment from its poetic employment. Elaborating on Aristotle's *Rhetoric* and *Poetics*, Ricoeur recognized a profound difference between the persuasive–cosmetic use of metaphor and its epistemological role in culture when used with creative, cathartic intentions. While the former is explicitly reference-based and defines rhetoric, the latter, which defines poetry, requires no direct connection or reference to the sensible world.

Just five years after Ricoeur's *The Rule of Metaphor*, Lakoff and Johnson published a book that focused chiefly on what Ricoeur originally defined as "the ontological function" of metaphor, its foremost formative role in shaping individual thought and framing social interaction. Lakoff and Johnson began their 1980 book by establishing their understanding of metaphor. Their fundamental definition closely resembles the concept of simulation as used in this book. For Lakoff and Johnson, "the essence of metaphor is understanding and experiencing one kind of thing in terms of another" (Lakoff and Johnson, 2003, 5 in Möring, 2012, 3). They took an experiential approach to understanding the metaphor as a powerful ontological (and even pre-ontological) instrument, they wrote: "no metaphor can ever be comprehended or even adequately represented independently of its experiential basis" (Lakoff and Johnson, 2003, 19). Lakoff and Johnson also specified that the ontological functionality of metaphors is always rooted in the bodily origins of any ontology, explaining that:

> Just as the basic experiences of human spatial orientation give rise to orientational metaphors, so our experiences with physical objects (especially our own bodies) provide the basis for an extraordinarily wide variety of ontological metaphors, that is, ways of viewing events, activities, emotions, ideas, *et cetera* as entities and substances. (Lakoff and Johnson, 2003, 25)

Closely resonant with both the postphenomenological perspective on the philosophy of technology adopted in this book and with Lakoff and

Johnson's recognition of the bodily origin of the ontological metaphors employed by humans, Alison McMahan has proposed an understanding of embodiment and technology according to which the body is itself an ideological category, a cognitive instrument (a medium) that precedes all other sociocultural influences and determinants (McMahan, 2003). On a similar note, but from a more general, hermeneutical point of view, De Mul observed that the semantic value of any kind of information always depends on the experiential horizon of the recipient (De Mul, 1999, 81). In the theoretical framework that I propose in this book and disclose with the videogames I design, the poetic, persuasive, and ontological effects of virtual worlds are understood as emerging by means of experiential analogies. Such analogies always originate and develop on the basis of a bodily (and thus positional) experiential horizon.

To be more precise in the definition of the experiential analogies encountered in virtual worlds, rather than thinking in terms of metaphors, it is useful to think in terms of allegories (or extensive metaphors). The term allegory indicates that the metaphorical scene put in front of our eyes has a more encompassing nature (as in an interconnected system of analogies within the representational space) and/or a certain duration and certain developments in time, which set it apart from the simple metaphor that is, by definition, instantaneous in its analogic function. A metaphor is the immediately mediated transferral of a unit of meaning from a literal context to a figurative one. As such, a metaphor is not characterized by development in time. Differently from metaphors, allegories can be characterized by a temporal dimension and by changes that occur during their enactment (e.g., in religious rituals, carnival parades, or theatrical representations). Allegories can also be characterized by a systemic interconnection of meaning among multiple metaphorical elements.

In their 2007 book *Gamer Theory*, Alexander R Galloway and McKenzie Wark proposed an allegorical understanding of the various relationships that take place between the elements that constitute virtual worlds or, more specifically, videogame worlds. In pursuing this purpose, they introduced the concept of allegorithm: "The gamer discovers a relationship between appearances and algorithm, ... that's the allegorithm" (Wark, 2007, 31). Wark further specified that:

> The allegorithm by which the gamer relates to the algorithm produces a quite particular allegory by which gamer and algorithm together relate to gamespace. In a game, any character, any object, any relationship can be given a value, and that value can be discovered.... [A] world

> in which any value is arbitrary, yet its value and its relation to other values can be discovered through trial and error. (Wark, 2007, 31)

Understood as allegories, virtual worlds can be interpreted as materializing their messages, ideologies, and worldviews by making them objectively accessible to their players by means of aesthetic stimuli and feedback devices (their graphical appearance, their sound effects, the vibrations of their controllers, etc.).

In proposing an understanding of virtual worlds as digital, poetic allegories I consider it not only interesting but also necessary, following Ricoeur, to investigate the cultural and cognitive effects that virtual worlds can engender, and to determine the kinds of social objectives we can aspire to achieve through their design, exploration and manipulation. A proceduralist perspective on the sociocultural role of digitally mediated simulations would contend that they can be used for a number of cultural scopes in a broad variety of social contexts, but that their ultimate function is, as already outlined, rhetorical. In other words, this perspective embraces virtual worlds as persuasive media whose influence on human beings can affect individual preferences, behaviors, and ideologies. The broader philosophical perspective advanced in this chapter was motivated by the recognition that the dominant (and largely proceduralist) theoretical horizon might be incomplete or even inaccurate. To encompass the way meaning can emerge in virtual worlds and to better structure an experiential framework to understand the ontological consequences and experiential potential of virtual worlds, I propose understanding virtual worlds not only as interactive, rhetorical allegories, but also (and perhaps mainly) as interactive, poetic allegories.

Similar to Ricoeur's understanding of the work of the poet as depending "not on descriptive truth but on conformity to his hypothetical postulates," game designers also conjure virtual worlds that have no necessary connection with the world they experience in their everyday existence. Both game designers and poets present worlds "as in act." The most striking difference between the two creative pursuits is that the medium through which game designers express and give shape to their vision allows for the emergence of worlds that can be experienced objectively and interactively. In this theoretical framework, to play indicates precisely the act of willingly performing under the acceptance of an allegorical detour.[2] This detour comprises setting up an artificial world that can be understood, explored, and manipulated on the basis of its analogy with the world to which human beings are native. These analogies are pursued in virtual worlds, as game designers and world-builders, through the same logic that binds literal meanings to figurative ones in literary allegories.

The mutually constitutive relationship between play and poetry was originally articulated by Johan Huizinga in his *Homo Ludens*, where, in the chapter on "Play and Poetry," he considered aspects that are common to both activities. "The affinity between poetry and play," Huizinga wrote, "is also apparent in the structure of the creative imagination itself. In the turning of a poetic phrase, the development of a motif, the expression of a mood, there is always a play-element at work" (Huizinga, 1955, 132).

Many of the core arguments presented in this book were inspired by Heidegger's philosophy of technology or are direct re-thematizations of his larger philosophical project in the age of digital mediation. It is interesting to observe, with the purpose of exploring the perspective that the design of virtual worlds has analogies with the poetic use of language, that on various occasions Heidegger presented an understanding of language as a fundamental tool to cognitively construct the world and operate on it. The philosophical perspectives and the logical arguments that led Heidegger to recognize the derivative essence of technology in relation to the deeper metaphysical context of Western thought in *Being and Time* are also the foundation for his attribution of the direct dependence of metaphorical language on metaphysical thought in *Der Satz vom Grund* (*The Principle of Reason*). In these writings, Heidegger explained that "the metaphorical exists only within the bounds of the metaphysical" (Heidegger, *Der Satz vom Grund*, 89 in Ricoeur, 2008, 282).

Heidegger's work was also strongly influential for Ricoeur. According to Ricoeur, it was an understanding of the role and limitations of the metaphorical order analogous to Heidegger's that motivated Aristotle to designate clearly understandable language as the only context in which metaphors could be expressed with clarity and understood (Ricoeur, 2008, 40, 41). It follows from this standpoint that poetry should mostly resort to common language to evoke the desired emotional and experiential effects more vividly. This expressive strategy can also find applications in forms that are non-textual and not strictly linguistic. For example, the various staged representations that pertain to theatre, myth, and religion all embed their metaphors in settings and circumstances with strong similarities or a direct connection to situations and occurrences that are frequently experienced by their intended audiences. This inherent and functional tendency toward immediacy in metaphorical and allegorical expression is commonly referred to as *mimesis* and was first employed in the context of Western philosophy by Socrates.

Regardless of its negative (e.g., those associated with sophistry, in the case of Socrates or Plato) or positive (e.g., as a strategy to elicit empathy and pursue catharsis and truth, as in the case of Aristotle's understanding of the social role of tragedy) undertones, the concept of *mimesis* traditionally entailed the imitation of some thing or situation as originally encountered, or as it is likely to be experienced, in the actual world. This interpretation implies that *mimesis* is a quality that is always developed and expressed within the horizon of certain ways of being-in-the-world (ways that are, ideally, largely shared by both the authors and the recipients of the contents). Hermeneutically speaking, the semantic value of any kind of information always depends on the experiential horizon of the recipient (De Mul, 1999, 81).

In sum, when trying to communicate or evoke human experiences and feelings, poetic language – or, more generally, any form of poetic expression – appears to be functionally characterized by three qualities:

1. The framing of the intended poetic effects within the cognitive horizon of the intended recipients (without which poetry would be unintelligible and would not lead to the intended experiential results).
2. The establishment of an accessible analogy between the intended poetic effects and the everyday, proximal experience (or lifeworld) of the intended recipients (without which poetry would be impossible to relate to and is likely not to be as immersive or captivating as intended).
3. The capability of evoking and disclosing new experiences (either fictionally or by means of simulation) that are unfamiliar or alternative to the phenomenological lifeworld of the intended recipients of the poetic expression (without which poetry would simply reference that world and, thus, cease to be poetic by definition).

In the poetic use of language, the referential fields of metaphors and allegories liminally extend beyond those of the sensible world and construct relationships that can be non-referential. The methods and logics of employment of poetic metaphors do not change when trying to afford feelings and experiences that are not phenomenologically comparable with how human beings experience and interact with the world in their everyday lives. Bogost claims, as did Nagel before him, that the subjective, embodied qualities of phenomenologies that are not human can never be fully and objectively retrieved. If that is the case, then mimetic approximation is not only the strategy to effectively mediate familiar feelings or experiences but also the conceptual instrument that can reveal unworldly ones. "In a literal sense," Bogost wrote, "the only way to perform alien phenomenology is by analogy" (Bogost, 2012, 64).

To summarize the core differences between ontology and metaphysics as espoused in this book, ontology is understood as a rationalization of a particular worldview, a certain relationship established by a being with reality, whereas metaphysics indicates a specific family of worldviews that follow from the establishment of a theoretical standpoint, a perspective based on a tradition of thought that methodologically separates the object of observation from an (abstract, ahistorical, and disembodied) observing subject. Disclosing an ontology other than metaphysics is the fundamental purpose of poetic metaphors.

Embracing this understanding of the sociocultural role of poetry, I do not believe that virtual worlds could be thoroughly or accurately understood exclusively as rhetorical instruments. The projectual qualities of computer simulations – observed, for example, in some of the philosophical uses discussed in the previous chapter – and their affordances for interactively materializing thought experiments, unconventional worlds, and alternative ways to experience things such as space, time, causation, and material properties, manifestly challenge the specific focus on *mimesis* that characterizes both the rhetorical and the ideological forms of (mediated) expression. In light of these reflections, I articulate in the following sections an understanding of virtual worlds capable of encompassing both their mimetic qualities and their poetic ones.

7.4 Metaphors and models

In its scientific connotation, a model is a heuristic instrument that serves to test, explain, and explore a certain interpretation of a phenomenon or set of phenomena (Black, 1962, 219–243). Disclosing new worlds and worldviews, the uses of poetic language, or poetic forms of expression more generally, can also be understood as heuristic models. This perspective closely aligns with the formative role of poetry in culture as presented in Heidegger's later work. Heidegger interpreted the work of art as a method to "open up" new worlds and worldviews and to mold culture, rather than as a passive by-product of a certain historical context. In Heidegger's later thought, the potential to disclose and establish new worlds, and new possibilities of relating to them, was extended beyond the work of art to embrace a wider spectrum of poetic forms of expression (Heidegger, 1982, 12, 13; Heidegger, 2000, 174).

The idea of an analogy – or even an identification – between heuristic models and poetic metaphors (or, more aptly, poetic allegories) was originally proposed by British-American philosopher Max Black. In his work Black identifies in isomorphism, a comprehensive structural correspondence between two systems, the logics underlying both the functional relationship between a model and the world to which the model refers and the poetic correspondence between an allegory and the world that the allegory references. Based on this isomorphic association, the poetic employment of *mimesis* can take the role of the foundation for the heuristic potential of human expression and can transcend the derogative Platonic understanding of art as a copy of second order. It is precisely in their projectual and heuristic employment that metaphors and allegories can effectively become cultural instruments instead of mere linguistic devices.

Inspired by Heidegger's philosophy in relation to technology, metaphysics, and poetry, I argue that virtual technologies are characterized by a meta-linguistic potential that transcends the limited possibility to influence human thought persuasively (meaning at the level of ideology), since they have the heuristic potential to operate at a deeper, ontological level. To reiterate, digitally mediated simulations can only be seen as having heuristic value and ontological effects if they are accessed as interactive, persistent worlds. Relying on a similar interpretation, Michael Nitsche attributed to virtual worlds the capability to fundamentally impact the way humans structure their thought, and he identified videogames as "a way of comprehending space, time and causation" (Nitsche, 2009, 43).

In the second half of this book, two claims are made for interactive experiences of virtual worlds to be understood as factors of sociocultural change. The first claim, mostly structured in Chapters 4 and 5, argued that interactive digital simulations can be used as ontological instruments and as mediators of philosophical thought. In other words, I recognized computers as media capable of overcoming the constraints and the effects of written text as the dominant form of mediation for the development and the dissemination of thought.

The second claim, presented in this chapter, attributes a more encompassing cultural role to interaction with virtual worlds than the didactic, entertaining, and ideological functions that are already solidly established in Western societies. More specifically, in this second claim I propose a vision of digital simulations as heuristic allegories, which is to say, devices that can operate on human cognition at a fundamental, ontological level.

Developers and designers of digital simulations must, however, be wary. As has also been argued by Marshall McLuhan, various forms of mediation have already proved their intrinsic potential for leading to profound psychophysiological changes in human beings. Changes that are often only fully understood after a medium has acquired a certain technological momentum and is already influentially integrated in social practices. Additionally, I would like to clarify once again that neither of the claims presented above intend to qualify interactive virtual worlds as neutral, as the definitive philosophical medium, or as the ultimate epistemological domain.

7.5 Mimesis and the "*irrealis*"

According to the anthropological perspective proposed by Plessner, the ambiguity and lacking that characterize the existential condition of mankind are the essential motivations behind the innate tendency of human beings both to complement their bodies through the creation of technical artifacts and to establish artificial worlds in the realm of culture. This refers to the same irreparably broken condition experienced by the positionally eccentric beings presented in Plessner's 1928 text, *The Levels of the Organic and Man*. Plessner labelled this defining characteristic "the focus on the *irrealis*" (Plessner, 2006, 334). This focus on the *irrealis* refers to human beings' propensity to try to overcome their contextual and situational limitations. These limitations fundamentally originate from mankind being bound to experience only the world labeled as actual, to experience this world exclusively as human beings, and thus to forever construct themselves in this relation to this double bind. In other words, and in a way that is reminiscent of the concept of the *fantastik* as explained by Novalis,[3] Plessner's focus on the *irrealis* reflects the existential need inherent to eccentric beings to escape, complement and constantly reconstruct themselves. The overt objective of this characteristic human propensity is to offer the individual temporary freedom from the seriousness and the boundaries of one's own identity, physical capabilities, and perceptual, emotional, intellectual, critical, and contextual thrownness.

This fundamental human necessity was also recorded, in its fundamentally bodily dimension, by Mark Hansen who (as already observed in Footnote 8 of Chapter 3) defined the concept of virtuality as "that capacity, so fundamental to human existence, to be in excess of one's actual state" (Hansen, 2003, 51). According to Hansen, the virtual is not "an abstract, disembodied dimension" but a "creative dimension of human embodiment itself – an excess of the body over itself" (Hansen, 2003, 90). This excess, the attraction to something other than one's ordinary engagement with the world (a drive that motivates activities such as daydreaming, playing, being absorbed by fiction, designing videogames, etc.), was considered by Huizinga as having such a fundamental role in social processes that he appointed it as the fundamental premise to any form of culture (Huizinga, 1955). In analogy with the core idea presented in this chapter, according to which poetic expression has a cognitive, heuristic function in culture, Huizinga asserted that

> *poiesis*, in fact, is a play-function. It proceeds within the play-ground of the mind, in a world of its own which the mind creates for it. There, things have a different physiognomy from the one they wear in "ordinary life," and are bound by ties other than those of logic and causality. (Huizinga, 1955, 199)

Following these observations, it is useful to begin an exploration of the ontological consequences of the experience of virtual worlds as emerging from three qualities of mediated content. These three qualities correspond to the three functions that, to different degrees and in different combinations, define poetic forms of expression, as discussed earlier in this chapter. These aspects are:

- **Intelligibility:** if a simulation were to take place in a logical–aesthetic context that were sensorily non-perceivable or indecipherable in the way it behaved or responded to user action, the experience of its interactive world would be trivially received, ultimately rendering the simulation ineffective.
- ***Mimesis*:** In the more general context of media studies, *mimesis* indicates the tendency of mediated content to be worldly in its appearance and behavior. By this I mean the tendency for mediated content to be structured and presented in isomorphic analogy to the way the actual world is experienced by the intended recipients of the said content. With a degree of *mimesis*, the mediated experience becomes more relatable to and can be grasped effortlessly and intuitively. It must be added that the concept of *mimesis* is neither absolute nor ahistorical, but is a quality of mediated content that changes together with sociocultural processes and aligns with tendencies and conventions, including those experientially established in virtual worlds. For example, in the context of game studies, the term *mimesis* is often employed to refer to the internal consistency of an interactive, digital

world. This specific quality, according to Roger Giner-Sorolla, also involves the thematic integration of logical puzzles and the contextual qualities of elements of gameplay within the overarching allegory of the game world (Giner-Sorolla, 1996).

- **Poetic unworldliness:** This third dimension of simulated content reflects precisely what was introduced above as the focus on the *irrealis*. It refers to the capability of media forms to evoke (in the case of traditional media) or materialize (in the case of digital simulations) worlds that are independent and often logically and aesthetically incongruous with the ones that human beings encounter in the context of their presence in, and biological dependence on, the actual world.

In the perspective proposed here, all three of these aspects need to be encountered and experienced together for an activity to be considered poetic from the point of view of a recipient. A practical example that is often used to illustrate the necessary involvement of all of the above-mentioned qualities in experiencing the poetic qualities of digital mediation is the analysis of a word processor (an interactive computer application used for the production, manipulation, and printing of texts).[4]

7.6 Poetically-simulated typewriters

The premise of my analysis of the role of a word processor as a mediator of human thought and experience is that a word processor is the digital remediation of the typewriter. A word processor can be seen as integrating actual, worldly functions of a typewriter with unworldly ones that are uniquely possible when mediated by the interactive and combinatorial affordances of computers. In their early manifestations, it was not uncommon for digital text editors to feature animations and sound effects that had the aesthetic goal of rendering the operations involved in the production of text (e.g., striking keys or starting a new line after a full stop) experientially analogous to that of a mechanical typewriter, and thus more familiar and intuitively comprehensible. The central element of the interface of text editors is the digital simulation of paper sheets. The simulated sheets frame the field of combinatorial, editorial possibilities of the text editor while imitating the aesthetics of earlier, more physical forms of text production. These analogies, again, are motivated by the intention of making the typing process easy to interpret and immediate in its basic functional behavior for all users, including those who are unfamiliar with the logics and behaviors that characterize the digital medium. The digital writing space's whiteness, proportions, orientation, and size in relation to the user and to the typographical symbols that can digitally appear on them are conspicuous examples of *mimesis* between traditional practices and their computer remediations.

Despite the many mimetic aspects that bind a word editor to the original functions of a typewriter, word processors are not limited in their writing and editing possibilities to those of a traditional, physical typing machine. They offer additional manipulative functions that rely on the affordances granted by the digital medium. These additional possibilities are frequently summarized in colloquial contexts as the ABCD of database ontology (indicating the basic computer operations of Adding, Browsing, Copying, and Deleting). Instead of focusing on each of the functions that is characteristically possible within the digital medium, I find it more relevant, and perhaps more intuitive, to focus at this point in the development of my argument on the ontological implications of three common possibilities offered by text-editing software. Albeit recognizing a degree of intelligibility as a fundamental quality of any poetic experience, in presenting the three examples below, I do not discuss the familiarity and the intuitive quality of the engagement with the world disclosed by a word processor. I accept those aspects, instead, as the pre-conditions for the (intended) meaning of a piece of software to emerge, and consequently for the actualization of both the mimetic and the unworldly qualities of the experience.

The three common affordances of text-editing software that I would like to explore in the subsequent sections of my text are: the undo function; the possibilities for cutting, copying, and pasting text; and the capability for importing and embedding several different kinds of digital information within files.

The undo function

The undo function of a word-editor allows users who are in the process of editing and formatting text to move back and forth in the operational time of their interaction with the software. With the undo feature, text can be manipulated in a way that is radically different from what would be possible on a physical sheet of paper with an actual typewriter and a physical ink-ribbon. In contrast to the principle of causation, which essentially defines every aspect of the way human beings are in the world, the causation that is simulated within the virtual world of a text editor is neither unidirectional nor inexorable. This new and very practical relationship with time that is revealed by the text-editing software requires its users to overcome their original, worldly understanding of time and causation that was originally shaped in an exclusive relationship with the actual world. The objective, pragmatic experiences of causation as multidirectional, and of the less prescriptive connections between phenomena in a (virtual) world, were not accessible to humans before the diffusion of interactive, digital mediation.

Cutting, copying, and pasting data

The operations corresponding to the cut, copy, and paste possibilities have an intuitive mimetic association with physically moving objects in the actual world. Moreover, they have a direct functional association with the traditional practice of manuscript revision, whereby human editors would physically cut paragraphs from one page of a manuscript and literally paste them onto another page. In text-editing software, the affordances for cutting, copying, and pasting allow text to be transferred from one location in a digital document to a different one within the same or a different document. Despite maintaining clear and strong connections with their worldly origin, certain possibilities for action offered by a word-editor transcend the materiality of the pre-digitally established way that text was manipulated. The possibility to infinitely cut, copy, and paste data defies the traditional understanding of space in the same way the undo function infringes upon that of time. Understood in this way, word-editors implicitly ask their users to relate to the virtual sheet of paper in front of them not as if it were physically concrete, but rather as a discrete aggregation of modular elements whose order and interconnection are never definitive.

In presenting an understanding of the affordances of cut, copy, and paste as defying the traditional understanding of space, it is relevant to note also that the potentially unlimited replication of text, and of digital paper sheets, is highly incongruous with the finite possibilities for operating with physical support for text, or in the physical world in general. Similarly to Borges' 1939 essay, "The Total Library," word-editors suggest that the (actual) world is nothing but a single instance among all the possible (virtual) combinations of its constituent elements.

Importing and embedding

Finally, I would like to consider the ontological effects of the possibilities offered by text editors for importing and embedding several different kinds of digital information in files. These affordances (for importing and embedding a variety of data forms) are not a unique feature of text editors. Rather, these are common possibilities of action in several digital platform applications. Ontologically speaking, every computer software application can be identified as structured on the fundamental assumption that everything within a virtual world, at its most basic level, consists of nothing more than digital information. This basic ontological approach is also foundational for the flow of content across multiple media platforms. With specific regard to the virtual remediation of traditional media forms, the ontological flattening (the perspective according to which everything that exists is ultimately made of the same substance and has the same status and value) also manifests itself in the shape of information convergence. Text ceases to be text; images, music,and other types of content adopt the form of discreet, modular, and configurable units in an indexable and transformable set of homologous data.

In a well-known passage of Borges' 1942 essay, "The Analytical Language of John Wilkins," the bizarre and unthinkable animal classification encountered in "a certain Chinese encyclopedia" featured a category for beasts that are "drawn with a very fine camelhair brush" (Borges, 2001, 231). In an analogous way, working within the flattening perspective outlined above, virtual worlds disclose experiences in which the difference that humans customarily perceive between an object and its (re)presentation is arguably less clear and less significant than within actual experiences.

7.7 Concluding remarks

In summary, this chapter has discussed the characteristics of digitally simulated worlds in terms of their mimetic qualities (which were recognized as responsible for the emergence and efficiency of their allegorical meanings), and in terms of their viability for affording poetic experiences (which is identified as the ultimate purpose of computer simulations when embraced as ontological instruments). Crucially, in both their mimetic and their poetic dimensions, the experiences emerging in relation to digitally mediated simulations are not only reliant on the technical affordances of computers. These experiences need also to be recognized as designed, encountered, interpreted, and manipulated based on the fundamental anthropological structures of human beings, on their cognitive capabilities and their biological thrownness.

In *The Levels of the Organic and Man*, Plessner noted that the effects of technology are, in all circumstances, largely outside the complete control and understanding of the technology's creators. I am convinced that the same is true concerning the changes and influences that the digital medium is producing in sociocultural processes in its role as a mediator of human thought and experience. As observed at the beginning of this chapter, Postman considered all forms of mediation to inherently impose assumptions and values on individuals and societies alike. Moreover, as McLuhan elucidated, society is in a constant, narcissistic state of technological hypnosis and is unaware of the effects and influences inherently enforced by media (McLuhan, 1994, 11). The technological assumptions and worldviews imposed by a technological environment only become partially recognizable after a certain critical or temporal distance is established between a society and a socially-embedded technology. Additionally, I argue that part of the thought-shaping and behavior-modeling influence that media and technology have on individuals and societies cannot be fully controlled by their human creators precisely because its impact relies upon and is driven by aspects of what it is like to be a human being. Aspects that are *themselves* outside the range of human control or even full comprehension.

8

Virtual Worlds and the Human Condition: Cognitive, Perceptual, Critical, and Operational Limitations

This book has articulated a vision in which digital simulations can be understood as mediators capable of granting human beings access to artificial worlds. This specific affordance offers us, for the first time in history, the possibility of developing human kinds of ontologies in ways that do not exclusively emerge from our relationships with the actual world. Through the mediation of videogames and digital simulations, augmented ontologies can be structured in relation to worlds that are independent from, and often incongruent with, the one human beings share as biological creatures and depend on as organisms. Virtual worlds disclose interactive experiences and phenomenologies that are not simply actually present but effectively extend toward what is virtually possible. In this particular age of anthropological enhancement, humans no longer simply design their lives existentially, but they also do so biologically and experientially.

The first half of the book (Chapters 1 to 5) lays the groundwork and establishes the fundamentals of the basic argument. In Chapter 6, the general argument of the book is enriched with an anthropological perspective according to which virtual worlds are disclosed for their players and visitors through what are effectively new, virtual bodies. Chapter 7 complemented this perspective with an understanding of digitally mediated simulations as a poetic declination of metaphorism. The present chapter completes the presentation of human beings' capability to generate metaphors by explaining how every way in which human beings can express themselves projectually in social practices has a necessary and inescapable foundation in the specific way we are in worlds (and in the actual one to begin with) as human beings.

As anticipated at the beginning of this book, the growing integration of digital mediation in social practices, together with the more pervasive and accessible possibilities for overcoming traditional kinds of human ontologies through virtual technology, is not seen as a paradigm shift, but rather as a deepening of the human project and as a resignation to the inevitably limited scope of the human project itself. Accordingly, this book has not argued that digital simulations grant the possibility of overcoming traditional Western ontologies in the sense of a surpassing and a forgetting of our metaphysical past, but rather in the sense of a *Verwindung* as presented by Heidegger, that is to say, as a form of overcoming that is both a distortion and a repetition of human kinds of ontologies. The concept of augmented ontologies did not aspire to encapsulate a cultural revolution, but rather embraced what was recognized as an epistemological and evolutionary shift (thus involving both ontological and anthropological changes). As part of this shift, the new possibilities opened up by the digital medium should not be understood as the context in which a novel set of ontologies can find a new, stable grounding, but rather, to reiterate a key proposition made in this book, as the phenomenological and experiential context where a new humanism has already begun to arise.

Having framed my argument, in this final chapter I formulate synthetic answers and present concluding remarks in relation to the questions concerning digital technology outlined in the first chapter. To review, these questions aimed to:

1. provide an ontological account of what everything looks like to our digital-hammer-wielding culture (a hermeneutical perspective), and
2. explore the expressive limitations and possibilities inherent in using virtual worlds specifically as philosophical mediators (that is to say, as ways to materialize philosophical concepts, perspectives, and thought experiments) and as technologies of the self.

8.1 The expressive limitations and possibilities of virtual worlds as poetic allegories

The postphenomenological approach to virtual technologies proposed in this text, and the consequent understanding of digital simulations as mediators of human thought and experience, are academic endeavors that engage very fundamental assumptions of philosophical discourse.

Even without focusing specifically on topics like the origin of consciousness or the properties of the mind, this book could not avoid implicating and being implicated in questions that have traditionally been ascribed to the realm of the philosophy of mind, rather than remaining within those of media philosophy or the philosophy of technology.

Unsurprisingly, the merging of areas of common interest to media philosophy and the philosophy of technology with topics that were traditionally the exclusive domain of the field of philosophy of mind is deeper and more frequent when the philosophical questions addressed are more fundamental. Three basic positions characterized by inextricable overlaps between these disciplines can be identified:

1. Ontologies are rational constructs that develop on the basis of a certain experiential horizon. As such, ontologies always have a bodily origin (in terms of their perceptual and operative dimensions) and depend on contextual factors. The rationalizations of the experience of a world that constitute ontologies must also be recognized as performed through specific kinds of rationality (in terms of their cognitive dimensions). Clearly, as for any cognitive operation, rationality in human beings has its origins and its developmental context in an embodied relationship with reality. In other words, this rationality needs to be identified as a specifically human kind of rationality.
2. Humans are beings characterized by a structural eccentricity. As explained in Chapter 6, because of their characteristic eccentricity, human beings can understand the relationship with their bodies as a relationship. They are also able to objectify their bodies and to reflect on their finitude, their identity, and the fundamental brokenness of their existence. Through the work of Helmuth Plessner, the eccentric positional structure is recognized as motivating the need for striving to complement and overcome the uncertainty and the incompleteness that define the existential condition of human beings. As such, human eccentricity is understood as both the cognitive background and the fundamental stimulus for the development of the artificial worlds of technology and culture.
3. Through the mediation of virtual and tele-technologies, human beings have achieved the next evolutionary structure of their cognitive organization. This new structure goes beyond a positionality that is simply eccentric and, precisely because of its type of fragmentation, was labeled by Jos De Mul as poly(ec)centric positionality. In this additional form of their experiential structuring, human beings can critically reflect on the relationships that they establish with their

native bodies and the limiting influence of these bodies on traditional kinds of human ontologies. The present book can be approached as a reflection on the ontological consequences of having reached that stage of the human relationship with digital technology.

Derivations and thematic elaborations of these three fundamental positions made up the three subsequent interpretations that were proposed and elaborated upon from different perspectives in this book:

(1) Technology as a sensory, cognitive, and operational extension of human biology

According to the understanding of technology adopted in this book (inspired and guided by the work of philosophers and media theorists like Kapp, McLuhan, and Postman), any human artifact can be interpreted as always having both anthropological causes and anthropological effects. In particular, technology is identified as a mediator poised to overcome the difficulties and disturbances arising from physical separation among human beings as well as from their limited extension in time and space (Weibel, 1992, 75). In this sense, technology is recognized as extending and fragmenting the perceptual, cognitive, critical, and operational capabilities of human beings, as well as allowing them to access new worlds, interpretations, and forms of interaction and thought. From the same standpoint, mankind can be understood as both the creator of its technologies and their by-product.

(2) Digital simulations as allegories with a bodily foundation

Similarly to what was observed in David Hume's analysis of the imaginative capabilities of the human mind, the perspective offered in this book finds that the fundamental human experiences "with physical objects (especially our own bodies) provide the basis for an extraordinarily wide variety of ontological metaphors" (Lakoff and Johnson, 2003, 25). By definition, both metaphors and (extended metaphors) have the fundamental sociocultural objective of communicating and explaining human thoughts, experiences, and feelings. This objective of metaphorism is always pursued through the process of analogy, regardless of how it is mediated (literary, ritual, theatrical, videoludic, etc.). Both the expressive potential of digital allegories and the methods of employment of metaphorism are consistent, irrespective of whether they are intended to present an interpretation of an actual event or behavior, or disclose new worlds and unfamiliar experiences. Put somewhat more simply, the process of metaphorism can both argue a case for what the world is like and for what it could be, or even what the world should (or should not) be, in the case of its normative and propagandistic uses. As already explained, all processes of analogy rely primarily on certain background experiences of the actual world. This is true for the creators of the analogy and for those at the receiving end of metaphorism. Consequently, and in accordance with Plessner's first and second anthropological laws, digital simulations can be understood as anthropological mediators capable of disclosing new dimensions of who we are and of new ways of pursuing complementarity and wholeness. As a result, virtual worlds foster extensions, distortions, and fragmentations of an established experiential horizon that is always determined and developed on a bodily foundation.

(3) The experience of digitally mediated content as a bodily activity

The stimuli that the virtual bodies of cybernauts receive in their experience of interactive virtual worlds, as well as variations in their emotional states, have direct, detectable consequences on their actual bodies. Evidence of these consequences does not come only from firsthand experience in relation, for example, to intense sections of a videogame or to especially engaging aspects of a simulation, but is also confirmed by the growing scientific interest and industrial employment of biometric methodologies to assess and understand the effects of virtual technologies on humans and animals.[1]

As is already detectable in the three interpretations above, uncovering the expressive limitations and possibilities of virtual worlds as poetic allegories is linked, in my work, to the concept of augmented ontologies. To summarize what augmented ontologies are and how they answer many of the questions that motivated and structured this inquiry, I take the concept apart and summarize the core related arguments made in this book.

8.1.1 The mediating qualities of computers

Any exploration of the expressive and epistemological potential of virtual worlds must be grounded in the analysis of the qualities and limitations of the mediator itself; those fundamental worldviews and the simulative affordances that define how the digital medium stores, modifies, presents, and communicates information. In a similar fashion,

Sebastian Möring commenced his analysis of metaphorism by looking at the core processes of the practice of simulating. According to Möring, the message or worldview that is expressed through a simulation "does not only depend on the characteristics of the available model but also on its implementability into a simulator. The materiality of the medium which is used to run the simulation does have an effect on the aspects which are implementable and might reduce these elements again" (Möring, 2012, 9).

The interpretation offered by Möring resonates with the understandings of the defining affordances of the digital medium cited during the development of my arguments. In that respect, I believe it is useful to mention again Lev Manovich, who explained that the digital medium is, at its logical core, nothing but a machine sustaining and representing autonomous or semi-autonomous logical systems capable of manipulating modular information. The endlessly configurable logics of interactive digital content is, according to Manovich, largely incongruent with the causal, univocal, and essentially stable way that humans have traditionally experienced and understood the world through the mediation of their senses (Manovich, 2001). Similarly, Ian Bogost has asserted that the specific type of persuasion inherent in the digital medium "is tied to the core affordances of the computer: computers run processes, they execute calculations and rule-based symbolic manipulations" (Bogost, 2007, ix).

In the development of my arguments in this book, as well as in the videogame examples cited, the digital medium emerged as particularly suitable for materializing worldviews that suggest, or even openly foster, the combinatorial triviality of any form of meaning or message. Anything experienced through a digital simulation will, in fact, inevitably be filtered through the ontological core of digital computation. As is the case for any form of mediation or technological augmentation, computers are recognized as framing both thought and *praxis* in a specific way, which is both advantageous and limiting in its applicability. Consequently, messages and worldviews originally intended by designers and materialized in interactive worlds are constantly under the threat of being distorted and rendered frivolous by the procedural rhetoric, the very interactive and combinatorial way the computer manipulates and presents information. From this perspective, McLuhan's observation that "the medium is the message" (the interpretation according to which the message of any medium or technology is "the change of scale or pace or pattern that it introduces in human affairs") appears to be particularly accurate (McLuhan, 1994, 8).

8.1.2 Humans who calculate

An observation complementary to the aspect summarized in the previous section was offered by Massimiliano Cappuccio from a perspective that encompassed computer science and the philosophy of mind. Cappuccio traced the steps in the development of modern technical computation back to their origin, the pioneering work of mathematician and computer scientist Alan Turing. To obtain a comprehensive understanding of the cognitive and computational qualities of the human mind that led to the creation of the first computing machines, Cappuccio pursued a joint reading of two fundamental texts by Turing, his 1936 "On Computable Numbers, with an Application to the *Entscheidungsproblem*" and his "Computing Machinery and Intelligence," published in 1950. According to the author's reconstruction, Turing first

> observed, analyzed and deconstructed the human element, and just then – on the basis of that analysis – he could engineer a machine that would imitate and simulate all its movements.... Before even emerging in Turing's academic treatments, his machine was already an eidolon, the transfigured representation of a man working behind a desk. (Cappuccio, 2005, 99)

With evident similarities to Cappuccio's observations, mathematician Robin O Gandy, a student of Turing's, noted that the analysis originally pursued by the British forefather of computer science "makes no reference whatsoever to calculating machines. Turing machines appear as a result, as a codification, of his analysis of calculations by humans" (Gandy, 1988, 83, 84). For this reason, modern computers are direct derivations, elaborations, variations, and evolutions of the original computing machines developed by Turing. Accordingly, regardless of their absolute precision and the indefatigably repetitive cycles of their computation, computers still clearly manifest and retain the biological imprint of the human kinds of ontologies that inspired the first Turing machines. On a similar note, the present text often identifies computers as the materialization (both in the sense of their hardware components and in the aesthetic worlds that they disclose and grant interaction with) of specific declinations of rationality and of our inescapably human ways of understanding space, time, and causation.

In a 1992 biographical volume on Turing, mathematician Andrew Hodges focused his attention on one characteristic in particular that, according to his interpretation, computers structurally borrowed from human cognition: the quality of possessing an intentional cognitive disposition toward objects of intellectual understanding (symbols). Along the same lines, Cappuccio went as far as asserting that each element that composes the machine, from both a logical and mechanical standpoint, presents traits that are inescapably anthropomorphic. Wittgenstein, a colleague of Turing's in the Cambridge years, also expressed this awareness with great clarity, stating that: "Turing's machines. These machines are *humans* who calculate" (Shanker, 1987, 615).

In the field of the digital humanities, Willard McCarty noted that the use of computing in the humanities is an instrument inherently limited in its modelling capability. Without explicitly mentioning the structurally human origin and analogies between the operational processes of the digital medium with those of human cognition, McCarty recognized "the fundamental dependence of any computing system on an explicit, delimited conception of the world or 'model' of it" (McCarthy, 2005, 21).

In this second component of answering the questions concerning digital technology, the logics of computers as well as their interactive, computational, aesthetic, and representational affordances are identified as structurally deriving from the inevitably biological, bodily origin of human cognition and design. As such, the alien phenomenologies and the augmented ontologies disclosed by the digital medium cannot be viewed as something genuinely other than human kinds of ontologies, but rather as their extensions, distortions, and fragmentations.

8.1.3 Metaphorism as "overcoming"

In the previous two sections, I have summarized two frameworks employed throughout this book to understand the expressive possibilities and the (crucially ontological and, by derivation, anthropological, epistemological, and political) effects of the experience of virtual worlds through digital simulations. In the inherent ambiguity that characterizes the subject matter of this book, these two perspectives can be identified as both incompatible and, at the same time, complementary. In the first perspective, virtual worlds are understood as technologies capable of disclosing experiences which are profoundly incoherent with the ones that can emerge in relation to the world that humans perceive and experience in their everyday lives. In the second perspective, virtual worlds are understood as cognitive, perceptual, critical, and operational extensions of mankind – extensions that cannot amount to something radically different or fully independent from the way humans are in the actual world as biological creatures. At first glance, there is an irremediable incongruence between these two frameworks.

The supposed incompatibility between the two interpretations is reminiscent of the ambiguity inherent in Heidegger's concept of overcoming (*Verwindung*). In other words, augmented ontologies proceed from the embedding of interactive digital simulations in social practices. As such, these augmented ontologies can be recognized as both a departure from and an inescapable repetition of the biological and cultural heritage of mankind. In a way that closely resonates with the observations presented above, Bogost proposed a vision of metaphorism that "is necessarily anthropomorphic, and thus it challenges the metaphysician both to embrace and to yield the limits of humanity" (Bogost, 2012, 74).

I believe that, as creators of virtual worlds and simulated experiences, our cultural role is compatible with that of those engaging in other declinations of metaphorism, regardless of their fictional or simulational nature. We tamper and negotiate with what we understand as the limits of what it is like to be human. One of the fundamental answers to the questions concerning digital technologies is that the experience of virtual worlds is a particular form of metaphorism that allows people to understand and negotiate the limits of what it is like to be human in ways that go poetically and effectively beyond their native relationship with the actual world and yet are inextricably bound to that relationship.

8.1.4 Where do alien ontologies come from?

From a standpoint inspired by Heidegger's existential phenomenology, this section elaborates an answer to a central epistemological question regarding the possibility of overcoming the horizon of pre-digital ontologies. Such a question can be presented as an epistemological paradox.

The topic of the alleged paradox addressed in this section is the unworldliness that can be experienced in virtual worlds. I am referring here to the fact that the worlds experienced through digital simulations do not necessarily behave similarly to the actual one or have any causal dependence on it. The question I am asking here concerning the unworldliness of virtual worlds can be formulated as follows: How can game designers, or more generally authors and creators of interactive virtual worlds, ever conjure experiences and phenomenologies that are unworldly, given the fact that they do not have the possibility to experience unworldliness in their everyday relations with the actual world?

This question can be treated as a paradox in relation to the adopted definition of ontology (the rational organization of a specific group of relationships constituted between a being and a world). From that standpoint, it follows logically that a human being who has never experienced an unworldly behavior or occurrence cannot be expected to possess the experiential and intellectual equipment to think and design in terms of that specific perceptual, cognitive, critical, or operational way of being. In other words, in this era of virtual worlds and digital simulation, the proposed paradox of inquiry could perhaps better be rephrased as follows: If human beings are the source of unworldly ontologies, where do these unworldly ontologies come from (given that the human beings cannot experience these ontologies before disclosing them)?

This enquiry is reminiscent of the epistemological paradox that Socrates formulated with Meno, a prominent Thessalian who visited Athens, in the Socratic dialogue of the same name. Socrates' own paradox of inquiry has its premise in the observation that "[a] man cannot search either for what he knows or for what he does not know. He cannot search for what he knows – since he knows it, there is no need to search – nor for what he does not know, for he does not know what to look for" (Plato, *Meno*, 80e, Grube translation).

In several cases, the qualities of virtual worlds that do not align with the actual world that human beings experience and engage with daily have an evidently derivative nature. By that I mean to say that some events and behaviors that can be experienced interactively in virtual worlds are nothing but complex ideas in the Humean sense; combinations or alterations of qualities, elements, or behaviors that humans have already experienced in the world they inhabit as biological creatures that, in their new and unexpected configuration, no longer resemble the original experiences. Among such derivative, combinatorial aspects, I believe it would suffice to cite the already mentioned combinatorial nature of videogame creatures, the affordance for reversing time that is increasingly common in contemporary videogames, and such possibilities as being immortal, having multiple lives, being able to perform a double jump, and pausing time. All of these examples can be recognized as simple alterations, reversals, or recombinations of worldly possibilities. As such, those virtual experiences cannot be considered to be truly disclosing new ontological horizons. Heidegger himself, in his "Letter on Humanism," expressed the belief that "the reversal of a metaphysical statement remains a metaphysical statement" (Heidegger, 1982, 208).

Another way to verbalize the central question of this section is to ask whether virtual experiences could ever aspire to transcend the pre-digital ontological horizons of human beings. Are there virtual worlds that, differently from the examples cited in the previous paragraph, effectively disclose experiences that transcend the perceptual, cognitive, critical, and operational limitations inherent in how human beings engage the actual world in their everyday lives *and yet* objectively materialize them in a way that they become possible to perceive, understand, and manipulate within the horizon of human kinds of ontologies? My answer is, in principle, yes. I believe that virtual experiences can (and already) do just that, and, on the basis of the strategy they adopt to overcome traditional (pre-digital) human kinds of ontologies, they can be understood as belonging to one (or more) of the following three clusters:

(1) Videogames and simulations inspired by computer malfunctions

Experimental titles like *Glitchhiker* (De Gier, Ismail, Nijman, Barbosa Dijkstra, Muller, Veer, 2011, available online at http://www.glitchhiker.com), offer interactive experiences that are inspired and disrupted by what appear to be unexpected glitches in the programming or the functioning of the game's hardware. Inspired by the graphical distortions and logical inconsistencies and omissions that can be experienced in relation to actual videogame malfunctions, the game proposed an experience that was removed enough from the consistency of human logic and perceptual possibilities to provide an exhilaration with the otherness of its functioning without transcending into unintelligibility. Another exemplary game that follows similar estrangement strategies, and relies on what was originally the disquieting, malfunctioning behavior of the medium, is the pioneering meta-game *ROM CHECK FAIL!* (Jarrad "Farbs" Woods, 2008, available online at http://www.farbs.org/romcheckfail.php) or in the sinister first person exploration *Memory of a Broken Dimension* (Ezra "XRA" Hanson-White, 2012, available online at http://www.datatragedy.com/wipmoabd/). I believe it is crucial to notice, however, that software bugs, hardware malfunctions, or any form of undesired behavior, ontological subversion, or procedural content unexpectedly produced by computers cannot be understood as anything other than deformed, faulty instances of human kinds of worldviews; phenomena and aberrations that are specific to computable environments that materialize certain human understandings of rationality and logics.

(2) Virtual worlds that materialized worldviews or insights that were originally disclosed by what Heidegger would call other "forms of revealing"

I am referring here to the possibilities of virtual worlds to grant interactive experiences of behaviors and dimensions that were originally explored, modeled or envisaged in theoretical or fictional/representational ways by other disciplines (encompassing scientific, artistic, humanistic and design efforts). The experimental videogame, *A Slower Speed of Light* (MIT Lab, 2012, available online at http://www.freeindiegam.es/2013/02/a-slower-speed-of-light-mit-game-lab), is clearly inspired by the tenets of Einstein's theory of relativity and allows players to experience how the world looks and feels (in a physically accurate way) to an observer who is gradually approaching the speed of light. Similarly, videogames like Valve's *Portal* (2007) or Marc ten Bosch's *Miegakure* (in development), take inspiration from the world of theoretical physics and offer their players the chance to simulate events and behaviors that it is simply not possible to have disclosed to us experientially in our everyday lives. Specifically, *Portal* simulates what it would feel like to travel through wormholes that short-circuit the continuity of space (although maintaining momentum), while *Miegakure* allows the player to interact and manipulate the game world in four dimensions plus time. In terms of more strictly artistic scenes of disclosure, several videogames currently offer virtual worlds that are interactive interpretations or renditions of artistic worldviews and concepts that were originally developed for non-interactive media. A recent example of such a remediation process can be seen in the videogame, *Bientôt l'été* (Tale of Tales, 2012, http://tale-of-tales.com/bientotlete), which overtly tries to capture the work of French filmmaker and writer Marguerite Duras in a simulational and interactive media form. In the case of this second family of ways overcoming can take place in virtual worlds, the relatable distance from being a traditional is achieved by experientially disclosing to their users/players theoretical insights that cannot be experienced in our native relationship with the actual world because of limitations concerning things such as: our scale as human beings; the fragility of our biological bodies; our perceptual, cognitive, critical, and operational equipment; and our individual subjectivities.

(3) Virtual worlds that try to objectify and evoke the unworldliness of psychotropic experiences

This third family of virtual worlds defies traditional human kinds of ontologies by resorting to the logical and aesthetic strategies of abstraction, synaesthesia, and of nonsense. Games like the tunnel shooters *Dyad* (][Games Inc., 2012) and *REZ* (Q Entertainment & HexaDrive, 2001) and the experimental, independent first person puzzle videogame, *This is Infinity* (Jonathan "Cactus" Söderström,

2009), are conspicuous examples of the design strategy presented in this third category. Several of the most disturbing and irrational videoludic experiments by Söderström openly referenced their psychotropic origin in their titles (e.g., *Psychosomnium*, 2008, and *Mondo Medicals*, 2007).

As explained in the three points above, I believe that virtual worlds can support and objectify experiences and worldviews that, albeit that they can be understood by human beings, are removed from their proximal experience and to an extent that they effectively transcend traditional (pre-digital) human kinds of ontologies. My answer to the questions concerning digital technology was not, however, a definitive yes. It was a conditional yes, a yes that is only such if we are ready to accept that the clusters of infringements listed above are alien and incompatible enough with the way each of us is in the world as a human being. Ultimately, all of the experiences listed above are still human experiences, but experiences that prompt and assist us in transcending and relativizing pre-digital kinds of human ontologies. The ontologies that can be developed as human beings in virtual worlds overcome the ones that were traditionally established and developed in relation to the actual world in the Heideggerian sense of their acceptance-distortion.[2]

The answer to the question of how we can create something we have never seen or experienced directly also resonates with the perspective on knowledge that is articulated in the Socratic dialogue *Meno*. Crafting unworldly, interactive experiences can, in line with what Socrates upheld, be possible, in principle, also without having had prior worldly experience of the very behavior, event, or process one might wish to understand or disclose. This was the case, I argue, in the first two videoludic estrangement strategies outlined above, where the boundary between knowing something and not knowing it was never clear-cut and absolute, but was presented as a gradient that can be bridged with intuition and with processes of analogy. This reflection ultimately outlines a phenomenological account of inventiveness and lateral thinking.

8.2 Well, well, let's get on with it...

Generally speaking, when approaching technology from an anthropological perspective, the successive stages of development and integration of a technological system in society are conceived as external materializations of successive phases of the self-understanding of man (Coolen, 1992, 250–271). In the more specific context of an anthropological approach to digital media, the interactive experiences of virtual worlds have been identified in this book as one of the possible contexts where the overcoming of the foundational and binding context of being thrown into this world as humans can take place.

The epistemological and expressive potential in the relationships that humans can establish with virtual worlds (and consequently with new, virtual bodies through interactive digital simulations) redefine us as creators, explorers, actors, and care-takers of a multiplicity of worlds, regardless of their actual or virtual constitution. Consequently, the question concerning the uses and possibilities of any kind of technology can only be asked as a corollary of a more fundamental question concerning the human condition. Attempting to delineate the absolute boundaries of the possibilities inherent in technologies is tantamount to trying to define conclusively what it is like to be human. Similarly, from the standpoint of Plessner's philosophical anthropology, man can be understood as a being that is impossible to characterize in a definitive way. The problems involved in the definition of man are attributable, according to Plessner, to the fact that any frame of reference for a possible definition (scientific, religious, philosophical,etc.) is always itself a cultural product of mankind. In this sense, human beings are bound to be involved in the historical process, entailing the impossibility of providing a comfortable, stable, and certain answer to questions concerning their identities, their role in the world, and the meaning of their existence.

I find that the last observation also appropriately fits the question concerning the expressive and ontological potential of virtual worlds as poetic allegories. The opacity and ambiguity that were often encountered in exploring and discussing issues such as the introduction and establishment of technologies in social processes and practices, the nature of human kinds of ontologies, and the definition of poetry can be recognized as deriving from the inherent needs of mankind to express and develop itself in the historical domain of culture. All of the guiding questions asked above, including the one concerning the ontological effects of the experience of virtual worlds, are impossible to answer definitively precisely because they are all involved in the very process of culture. They are factors contributing to the shaping of the psychological, socio-political, technological, intellectual, and experiential contexts that human beings shape and are, at the same time, shaped by. As a consequence, the delineation of the limits and possibilities of the enhancement of human perceptual, cognitive, critical, and operational possibilities cannot be achieved except in answer to open questions (or questions that can only be answered in ways that are temporary, indicative, and largely incomplete).

Regarding the inevitably incompletable, self-referential, and ephemeral nature of culture that emerges through all forms of social product or intellectual endeavor, Plessner tersely noted that

> What becomes concrete in the sphere of culture clearly shows its being inescapably bound by human authorship and (in the same amount) by human independence. Man can only invent what he has already discovered. (Plessner, 2006, 344, my English translation from the Italian edition)

Analogous to Plessner's observation, I would like to conclude with a quote from Heidegger extracted from the basic text for *What is Metaphysics*? Heidegger's inaugural lecture at the University of Freiburg in 1929. In the introductory statements, Heidegger reminded his addressees that "every metaphysical question can be asked only in such a way that the questioner as such is present together with the question" (Heidegger, 2008, 93). In this sense, the questions concerning digital technology must be understood as having been asked and explored in ways that are unavoidably historical and human. From the perspective presented in this book, and even when armed with digital hammers, our efforts cannot ever aspire to transcend the (flexible) walls of our inescapable humanity. Resigned and unrequited like characters in an existential play by Sartre, we can only keep hammering away. Well, well, let's get on with it...[3]

Notes

1 The Questions Concerning Digital Technology

1. This meaning aligns with Aristotle's perspective presented in his writings on nature (*Physics*).
2. When referencing Heidegger's 1927 book, *Being and Time*, I will use a dual canon for citation. The first part of all parenthetical citations to *Being and Time* indicate the position of an idea or quotation in the 1962 English translation by John Macquarrie and Edward Robinson. In the case of direct quotations, the second part of the dual canon indicates where the same idea or quotation is to be found in the original 1927 German edition (*Sein und Zeit,* abbreviated to SZ).
3. From a postphenomenological perspective, for example, the activities of reading a book, being engaged in a conversation, or preparing a meal, cannot be suitably understood as relationships where a "subject" directs itself towards a certain "object" or set of objects (a book, some people, the ingredients for a dish). Instead, through framework of postphenomenology, one is always in an intricate network of relations with the world, and in those relationships subjects and objects are intertwined and give meaning to one another (Verbeek, 2011, 28).
4. According to the mechanistic worldview proposed by the Enlightenment, reality should be describable objectively through mathematics because it is itself encoded with a mathematical language. In this sense, the digital worldview is a continuation of the mechanistic one. Similarly, Heidegger claimed that the advent of the technological world marks the culmination of the project of Western metaphysics to dominate a world made of objects (Heidegger, 1982, 120–127; Richard Villa, 1996, 182).
5. Excerpt taken from Hölderlin's 1803 hymn, *Patmos*.

2 A Reflection on Metaphysical Thought and Its Technological "Overcoming"

1. Andronicus of Rhodes specifically labeled those writings "*τά μετά τά φυσικά βιβλία*" (*ta meta ta physika biblia*): the books that come after the ones on physics. The Catholic Encyclopedia of 1913 explains that Andronicus' reference was mistranslated by Latin scholiasts. Instead of being interpreted as a bibliographical indication, "… it was understood to refer to the chronological or pedagogical order among our philosophical studies," thus indicating the knowledge of what extends beyond the physical world. One of the connotations of the term metaphysics that will be discussed later in this chapter (which colloquially understands metaphysics as a discipline concerned with transcendental speculations that are frequently related to spiritualism and the afterlife) relies precisely on that scholastic misinterpretation.
2. An example of a radical ontological approach that will be particularly relevant to the pracademic efforts offered in this book can be found in George

Berkeley's immaterialism. Berkeley's perspective – and in particular his elaboration of subjective idealism – was in fact the key inspiration for a small, independent videogame I designed and developed in 2013 titled *Necessary Evil*. *Necessary Evil* and subjective idealism were not, however, simply connected by personal interest or creative inspiration. George Berkeley's immaterialism was, rather, the very worldview that I tried to disclose experientially for my players, who were invited (and expected) to grasp Berkeley's perspective experientially through "play," instead of having to rely on the subjective interpretation of philosophical texts or lectures. A detailed analysis and explanation of *Necessary Evil*, and how the virtual worlds of videogames and digital simulations can be used as philosophical mediators will be the focal points of Chapters 3 and 4.

3. Following Pyrrho, Descartes imagined that the sensations experienced during waking life cannot logically be given a different ontological value than those experienced in dreams, psychoses, or deceitful hallucinations, and can be considered, as such, entirely illusory. The "evil d(a)emon" (sometimes called the "evil genius") is a speculative concept specifically introduced in Descartes' 1641 *Meditationes de Prima Philosophia* to illustrate his position on the illusory nature of empirical knowledge. Hypothesizing the existence of an all-powerful deceiver capable of intervening in our sensory perception, it would be logically impossible, according to Descartes, to determine whether perceptions have any relation to a possibly existing reality. The idea of the evil genius is closely related to the idea of a "consensus reality" proposed by the already mentioned George Berkeley.
4. Among the arguments most frequently presented in opposition to Cartesian perspectives regarding the mind-body problem are the problem of other minds and the impossibility of conceiving the idea of causal interaction dualistically. Contentions against the possibility of accepting a stance inspired by Descartes' dichotomy that are less purely speculative were developed on the basis of physics (as in the argument according to which a non-physical mind violates the law of conservation of energy) and neuroscience (including arguments relating to brain damage formulated by Paul Churchland, according to which brain injuries have a direct impact on a subject's ability to memorize and elaborate information).
5. Plessner's understanding of both the limitations and the merits of modern dualism has a close affinity with the interpretation of Cartesian philosophy offered by the French philosopher and social theorist Michel Foucault. In his 1966 book, *The Order of Things*, Foucault noted that, before the Classical age, human perception of reality was metaphorically embraced as the language of God, as Berkeley puts it. In other words, knowledge was considered a system whose presence, stability, and correspondence with noumenal reality were metaphysically guaranteed by a higher, ineffable order (Foucault, 1994, 58). Before the seventeenth century, the decipherment of such a system, implying access to an absolute framework of knowledge with a theological grounding, was considered the sole task of culture.
6. In his treatise *Περι Ψυχης* (On the Soul), Aristotle explicitly compared the mind with an unscribed tablet, "What the mind thinks must be in it in the same sense as letters are on a tablet which bears no actual writing; this is just what happens in the case of the mind" (On the Soul, 3.4.430 a1).

7. Albeit ingenious and intuitively graspable, the metaphor of wearing tinted glasses has some problematic aspects in its correspondence with Kant's insights. Drawing an analogy between a pair of colored lenses and the imperfect and contingent qualities of human experience suggests, in fact, that human beings could eventually remove these glasses or have them removed. Kant, however, never addressed or suggested the possibility of overcoming our "forms of receptivity" in the *Critique of Pure Reason*. On the contrary, he considered that those filters and limitations characteristically defined the way human beings are innately in a relationship with reality.
8. The *noumenon* (from the ancient Greek word *νοούμενον*, present participle of *νοέω*: I think, I mean) is an object, a quality, or an event that is independent from the presence of an observer and from the mediation of the observer's senses. After the work of Immanuel Kant, *noumena* are understood as the epistemological goals of *a priori* knowledge. *Noumena* are customarily presented as standing out, in a theoretical separation, against the *phenomena*. The term *phenomenon* (deriving from the ancient Greek word *φαινόμενον*) indicates any object, quality, or occurrence that is the subject of sensory experience.
9. In his poetic fragments, Xenophanes rejected the representation of the gods that had been paradigmatically introduced in the classical Greek period by Homer and Hesiod. Xenophanes believed it to be an absurdity to confer anthropomorphic traits (both physical and behavioral) on the gods. Such identifications are inherent to the way human beings perceive and interpret the world, rather than pertaining to divinity or to the essence of natural phenomena. In one famous passage, Xenophanes ridiculed the idea by claiming that:

   ... if cattle or horses or lions had hands and could draw,
   And could sculpt like men, then the horses would draw their gods
   Like horses, and cattle like cattle; and each they would shape
   Bodies of gods in the likeness, each kind, of their own.
   (Diels & Kranz, 38–58)
10. "Regional ontologies" are, in the original formulation of Edmund Husserl (1859–1938), ontologies that are developed in particular sociocultural domains (e.g., in the contexts of genetic biochemistry, game-design, Christian angelic hierarchy, etc).

3 Worlds in the Age of Digital Simulation

1. The field of ACI (animal–computer interaction), for instance, is a branch of the digital humanities in which the academic and design efforts are not directly aimed at the satisfaction of human needs and often diverge (by design) from human kinds of perspectives (Westerlaken & Gualeni, 2014a; Westerlaken & Gualeni, 2014b). Several declinations and examples of speculative design that can be encountered in Dunne and Raby's 2013 book, *Speculative Everything: Design, Fiction and Social Dreaming*, also purposefully encourage non-human and/or unconventional alternatives to the established uses and customary perceptions of design objects and related social conventions.

2. The adjective ergodic derives from the ancient Greek words *ergon* (έργον), meaning work, and *hodos*, (ὁδός), meaning path. In *Cybertext: Perspectives on Ergodic Literature*, Aarseth offers one of the most commonly cited definitions of what it means for literature to have the quality of being ergodic: "In ergodic literature, nontrivial effort is required to allow the reader to traverse the text. If ergodic literature is to make sense as a concept, there must also be nonergodic literature, where the effort to traverse the text is trivial, with no extranoematic responsibilities placed on the reader except (for example) eye movement and the periodic or arbitrary turning of pages" (Aarseth, 1997, 1).
3. The artistic efforts of the French group *OuLiPo* (acronym for *Ouvroir de Littérature Potentielle*, Workshop for Potential Literature) to disengage literature from meaning, authorship, interpretation, and even prescribed ways to approach reading, can be seen as particularly obvious examples of a budding cultural shift toward modality and a simulational mindset.
4. Technological determinism supports the general view that changes in technology exert a greater influence on societies and their processes than any other factors of cultural change (Memo from Wade Roush to Merritt R. Smith, in Smith & Marx, 1994, 2). Determinists regard technology as an autonomous force that develops (and spreads) according to internal and inscrutable logic. Social constructivists, in contrast, argue that new technologies are always the material answer to some intangible social need. They emerge out of "conflict, difference and resistance" that human beings have experienced in the course of history in a variety of sociocultural settings (Bijker & Law, 1992, 11).
5. The lexical choice of the term context over the more commonly used space or place is intended to make the definition of world applicable to interactive experiences that might not feature any explicit spatial dimension, such as textual-adventure videogames, book games, card games, or paper-based role playing games.
6. An interesting definition of reality that also aligns with the tradition of metaphysical thought in pointing out how sensations and impressions are neither sufficient nor necessary conditions for determining the objectiveness of any phenomenon was provided by Philip K Dick in 1978. According to Dick, "[r]eality is that which, when you stop believing in it, doesn't go away" (Dick, 1978).
7. Quick-time events are interactive sections of videogames that are featured less and less frequently in contemporary titles. It is my understanding that their characteristic gameplay fell out of favor due to their often abstract nature, their removing the player's direct control over the playing characters, and the consequent difficulty for coherently and elegantly integrating them in the game world without damaging its behavioral consistency and the depth of the player's immersion.
8. It is relevant to mention, at this point, that in his 2003 book *New Philosophy for New Media*, Mark Hansen also advocated for a reformulation of our understanding of what we mean by virtual. In line with Lévy's perspectives, according to whom virtuality is a quality that is inherent to a being and its projectuality, Hansen proposed an interpretation of virtual as "that capacity, so fundamental to human existence, to be in excess of one's actual state" (Hansen, 2003, 51).

9. In the concluding statements of an article I wrote in 2012 for *CONTROL ONLINE*, the magazine for the Dutch videogame industry, I explained that an "ideology is a system of beliefs that constitute one's goals and determines one's actions (for example game design decisions). Ideologies are often ideas accepted and proposed by the dominant class of a society that are, then, shared by its members" (Gualeni, 2012b). Therefore, when crafting the logical system(s) that are present in a game, it is inevitable that several ideologies contribute to the choices made by the designers, who are often unaware that such beliefs have a fundamental role in their process. "If those ideologies are the most basic meaning of videogames, then the visual materialization of the game *Duck Hunt* is a statement of superiority of mankind over animals; killing animals for fun and training them for our amusement are portrayed by the game as something acceptable if not openly desirable. Similarly, what *Farmville* is offering to its players is fundamentally an extension of the routine and objectives of their jobs into what should be their leisure time. The ultimate meaning of *Farmville* is economic efficiency, an objective that the player is led to pursue by means of also treating people as production resources. *Zynga's* 'masterpiece' does not teach the players anything realistic or valuable about running a farm, instead it rehearses the very capitalistic vision that motivated its development in the first place" (Gualeni, 2012b).

4 Thinking with Virtual Worlds

1. This passage from Borges can be found in the essay "El Idioma Analítico de John Wilkins" ("The Analytical Language of John Wilkins") included in the collection, *Otras Inquisiciones* (*Other Inquisitions*), first published in 1952. The complete translated text can also be found in Borges, 2001, 229–232.
2. Such central features are particularly evident in the words of Romanian poet, essayist, and founder of the European Dada movement, Tristan Tzara, who concluded his famed "Dada Manifesto 1918" as follows: "The abolition of logic, which is the dance of those impotent to create: DADA; ... every object, all objects, sentiments, obscurities, apparitions and the precise clash of parallel lines are weapons for the fight: DADA; abolition of memory: DADA; abolition of archaeology: DADA; abolition of prophets: DADA; abolition of the future: DADA .... Freedom: DADA DADA DADA, a roaring of tense colours and interlacing of opposites and all contradictions, grotesques, inconsistencies: LIFE."
3. The general fascination of modernism with tribal art and the emergence of regressionism in social production, such as body art, land art, and arcology, are conspicuous examples of this second modern mythology at work.
4. Oskar Schlemmer's 1922 *Triadic Balled* (*Triadisches Ballett*) was inspired by the Christian idea of the Holy Trinity. It features three acts, three participants (two males and one female), 12 dancers, and 18 costumes. The costumes were essentially built with juxtaposed primitive geometric forms that made the dancers look like giant, abstract marionettes. Schlemmer considered the artificial precision of the movement of puppets and marionettes to be aesthetically superior to the way humans traditionally danced. Schlemmer

designed and produced many similar performance pieces, including his 1923 *Figural Cabinet*, described as a "mechanical cabaret," and the futuristic *Men in Space*. For more information and deeper insights into this topic, see Susanne Lahusen's 1986 article "Oskar Schlemmer: Mechanical Ballets?" (The complete text can be found in *Dance Research*, Autumn 1986, Vol. 4, Issue 2, 65).

5. The verb to describe derives from the Latin *scribere* (to write) and denotes, in its original meaning, the act of providing a written account of something. What Wittgenstein recognized as the primary role of language, that of mapping and appropriating the world by means of an artificial logical system, is apparent in this etymology.
6. The concept of momentum is used here with reference to Thomas P Hughes's theory of technological momentum presented in Chapter 3 of this book.
7. Clearly, this first understanding of what transformative experiences are is not meant to be interpreted deterministically. It should also be clear that transformative experiences are not the direct and predictable output of transformative activities, but instead emerge from the coactions of the transformative qualities of specific activities and the influence of individual sensitivity, previous experiences, and a broad assemblage of cultural determinants.
8. Pedercini demonstrated a clear awareness of the fundamental (and often overlooked) expressive role of the "unsaid" in transformative simulations and videogames. Details and aspects of a certain experience that are removed or simplified for the sake of a certain argument, as well as the boundaries of the game levels and virtual worlds (the "invisible walls and puffy clouds") that limit and guide players' behavior, clearly play a role in the "power" dimension of game design. Pedercini warned us that "[a]s designers we should be constantly aware of where we draw these boundaries, because it's easy to mistake scoping for a purely technical choice" (http://www.molleindustria.org/blog/invisible-walls-puffy-clouds/).

5 Augmented Ontologies and a Challenge to Western Philosophy: Videogames and Simulations as Mediators of Human Thought and Experience

1. The importance of transferring content with phenomenological immediacy – not in fictional and/or representational ways – was professed by nineteenth and twentieth century novelists, philosophers, playwrights, and movie directors who aligned with the existentialist current. The existentialist method embraced by Jean Paul Sartre purposefully aimed at transferring experiences, rather than ideas, to the reader. This objective was usually pursued through *mimesis*, by relying on the intended reader's existing familiarity with the situations and sensations described or represented (Sartre, 2010). For the reasons explained in the previous chapter, the textual medium (especially when utilized in a linear fashion) must be recognized as unsuitable for conveying experiences because of the opacity and limitations of its code, and its reliance on the reader's past experiences and capability for abstracting and imagining rather than producing immediate and objectively materialized situations. It is perhaps for this reason that Sartre himself relied on theatrical plays as well as books to spread his ideas, and that existentialism and expressionism found a particularly fecund soil in the cinematic medium.

2. On the official website for *Miegakure*, the author Marc ten Bosch notes that "[o]ur world is three-dimensional: width, depth, and height. But what if there was a fourth physical dimension that we cannot see, in addition to the other three? This game is about exploring the consequences of being able to move in four spatial dimensions + time. It plays like a regular three-dimensional platformer, but at the press of a button one of the dimensions is exchanged with the fourth dimension, allowing for four-dimensional movement" (http://marctenbosch.com/miegakure).
3. The design objective of materializing an alien phenomenology in the most transparent and immersive way was pursued in *Haerfest* with the use of the first-person perspective and the almost completely diegetic quality of the in-game music and sound effects. The relationship between the immersive quality of the experience of media content and the persuasive potential of computer applications has been the focus of studies and publications from a broad spectrum of academic perspectives (aesthetic, narrative, biometric, etc.). It is reasonable to think, through analogy, that qualities such as immediacy and transparency would also cognitively facilitate the formation of augmented ontologies.
4. For a more thorough discussion on the ironic and self-critical dimensions of *Necessary Evil*, see my gamasutra.com featured blog post entitled "Self-reflexive Videogames as Playable Critical Thought," available online at: http://www.gamasutra.com/blogs/StefanoGualeni/20131029/202847/SELFREFLEXIVE_VIDEO_GAMES_AS_PLAYABLE_CRITICAL_THOUGHT.php
5. From a game design perspective, it might be interesting to further observe that, like most games and videogames that take a critical stance toward socio-cultural processes or situations, *Necessary Evil* relies on controls, interaction norms, and aesthetic conventions that are already established in the tradition of a particular game genre (*Necessary Evil* specifically relies on the established conventions of the action-role-playing-videogame genre). The deliberate design decision of not pursuing design innovation and of relying, instead, on the repurposing of time-honored conventions has the double advantage of:
    1) not having to teach the players how to understand the world and operate in it, allowing them to access the critical message of the game in a more immediate and efficient way, and
    2) making the subversive, critical aspects of the game more evident by contrast; making them stand out as unexpected and unfamiliar against the backdrop of what can be considered largely already known by the players.
6. As revelatory examples of this approach, Mary Flanagan (2009) uses the term game as a synonym for play scenario, and, according to Ivan Mosca, game developers supply props to play with "like engineers supply technologies for flying and therapists supply tools for understanding ourselves" (Mosca, 2013, 19).

6 Positionality in the Digital Age: Virtual Bodies and the Effects of Virtual Experiences

1. In an oral culture, information necessary to the survival of a community or for the fostering of a certain set of social values and behaviors is passed from one generation to the next in rhythmic formulas rich with alliteration, memorable figures of speech, and mottos. The singing of an oral epic, for example, cannot

amount to the reproduction of a fixed text. There is no original version and no authorship. There is no correctness or incorrectness of recollection (Derrida, 1981, 134–142; Carr, 2010, 56). "He who thinks by speaking, learns by hearing, his thoughts do not belong to him, they belong to everyone. Homeric Greek has no words to represent mental events... [t]here is no vocabulary to express abstract cognitive states or processes." (Nyíri, 1993)

2. The Cartesian principle "I think, therefore I am" is ontological solipsism; it is a certainty that cannot be anchored to anything external to the mind of the thinking subject itself. This isolation is also apparent in "the presupposition that man as an autonomous subject must be guided solely by his own reason. As such the Cartesian concept of subject is part of the foundation of the modern-day liberal view of man" (De Mul, 2010, 167).
3. Philosophical anthropology is a philosophical position that understands humans as products of their biological, cultural, socio-political, and technical environments, as well as creators of their own value systems. In his 2012 paper "Philosophical Anthropology – A third way between Darwinism and Foucaultism," Joachim Fischer explained that philosophical anthropology "developed as a discipline through contributions from different contemporary paradigms, such as psychoanalysis, philosophical hermeneutics, existential philosophy, the phenomenology of the body, the phenomenology of human *Lebenswelt*" and so on (Fischer, 2013, 20).
4. This book does not propose that Plessner's theory of positionality be accepted literally and unquestioningly. For the sake of simplicity, in fact, the theory of positionality subdivides the organizational possibilities of organic life in a way that is discrete and rigidly compartmentalized. The various organizational possibilities could perhaps be better understood here as a gradient, instead. In this gradient, there is the possibility for certain animal species to be endowed positional structures that transcend their characteristic closed form. Such animals (for example, crows and many kinds of primates) are capable of crafting and using tools for functional, entertainment, and expressive purposes. In other words, some animals are capable of performing "eccentric" activities (including masturbating and recognizing themselves in mirrors) without having fully developed an eccentric positionality. In the positionality gradient, human infants and young children must also be identified as having a bodily organization that is not fully eccentric (or not yet eccentric), as they are incapable of performing tasks that require self-reflection and self-objectification.
5. The discrepancy between Plessner's and Heidegger's understandings of the duplicity inherent in being-in-the-world as human beings could be bridged by adopting an ancient Greek perspective on the relationship between the human condition and the passing of time. The Greeks, writes Robert M. Pirsig in the afterword to his 1984 *Zen and the Art of Motorcycle Maintenance: An Inquiry into Values*, "saw the future as something that came upon them from behind their backs with the past receding away before their eyes. When you think about it, that's a more accurate metaphor than our present one. Who really can face the future? All you can do is project from the past, even when the past shows that such projections are often wrong. And who really can forget the past? What else is there to know?" (Pirsig, 1984). The inconsistency in the position of a human being who is reflecting on him/herself from both "behind one's

back" and "ahead-of-oneself" in the linear progression of time can be overcome with the allegorical understanding reported by Pirsig. Interestingly, the later Plessner also adopted a perspective on human projectivity that is comparable to Heidegger's "being-ahead-of-oneself." According to Plessner, finding one's own equilibrium – a stable identity – cannot simply happen once and for all because the eccentric nature of man can never be overcome. For this reason, human beings have to perform recklessly and develop continuously toward new horizons and achievements (Boccignone, 2009, 6). To achieve a (temporary) balance they will always strive for something new, they will have to surpass their own deeds in an eternal process (Plessner, 1980–1984, Vol. IV, 395).

6. In Chapter 4, the understanding of the social role of art during the modernist period was observed through the lens of two defining ideologies. According to the first "modernist myth", the appreciation of a work of art has the potential to estrange people from their everyday context and means-end rationality, freeing their mind and facilitating the emergence and adoption of new (socio-cultural) perspectives. The second ideology, on the other hand, considers that the practical crafting of an (art) object or a product can lead the individuals involved in the creative process to the establishment of a more direct and genuine engagement with themselves and with the world (see Section 4.1 Pre-digital media forms and their ontological influence).
7. John Heartfield (1891–1968) is the Anglicisation of Helmut Herzfeld. The German artist decided to legally change his name in Heartfield in 1916 as a mocking statement against the anti-British attitude that was common in Germany during World War I.
8. The idea of ontological levelling and indifference to values as the ultimate meaning of a simulation or a combinatorial phenomenon was notably encapsulated in the infinite monkey theorem, first envisioned in 1913 by mathematician Émile Borel (1871–1956) with the purpose of demonstrating the relevance of timescales in statistical mechanics. The infinite monkey theorem affirms that a primate hitting keys at random on a typewriter keyboard for an infinite amount of time will "almost surely" recreate a particular chosen text (usually, exemplary works that hypothetical monkeys are supposed to type and belittle in derivative examples are William Shakespeare's tragedies). Here, "almost surely" is a mathematical indication with a precise meaning, denoting a non-zero possibility applied to an infinite sequence of events. In Borel's theorem, the monkey is a metaphor for an abstract device that produces a random sequence of letters *ad infinitum*. Borges adds to Borel's vision that "Strictly speaking, one immortal monkey would suffice" (Borges, 2001, 215).

7 Virtual Worlds as Poetic Allegories

1. In relation to Heidegger's criticism, and as already outlined in the previous chapter, both Ong and Havelock supported the thesis according to which Plato's Doctrine of Ideas can be interpreted as a direct consequence of the adoption of the textual medium in Greek culture between the sixth and fourth centuries BC.

2. In this sense, the activity of play can be understood to be a form of adaptation, only not directed toward the world that we inhabit as biological organisms. In a way that resonates with philosopher and sociologist George Herbert Mead, the activity of play can be described as a willing adaptation to an environment that is not there (Mead, 2001).
3. For the Romantic author and philosopher Novalis (Georg Philipp Friedrich Freiherr von Hardenberg, 1772–1801), the concept of the *fantastik* refers to the mental faculty of free association of thought and to the poetic use of that same faculty.
4. Similar dissections can be found, for example, in Don Ihde's *Technology and the Lifeworld* (1990) and in the 1993 paper "Thinking with a Word Processor" by Kristóf J Nyíri. A comparable analysis was carried out by Jay D Bolter in his books, *Remediation–understanding new media* (2000, with Richard Grusin) and *Writing Space: Computers, Hypertext, and the Remediation of Print* (2001).

8 Virtual Worlds and the Human Condition: Cognitive, Perceptual, Critical, and Operational Limitations

1. The possibility of analyzing and assessing virtual experiences by means of psychophysiological experiments with creatures other than human beings is demonstrated by a growing number of studies relying on biometric observations in the recently founded field of academic research labeled Animal Computer Interaction (ACI). As a byproduct of the main body of work developed at NHTV Breda University of Applied Sciences in the field of biometrics and user experience research, Michelle Westerlaken and I have collaborated on a few papers in which we propose a novel theoretical foundation for ACI that relies on Plessner's theory of positionality. Our papers advocate the use of biometrics and grounded theory in combination with traditional observation of the animal's indexical signs during play as the phenomenological foundation for the development of a more objective and zoo-centric understanding of the animal's needs and preferences.
2. As explained in the introductory chapter and rehashed here, the process of overcoming is interpreted in accordance with Heidegger's interpretation of the concept. Accordingly, the term overcoming is not understood in the dialectical meaning inherent in the German term *Überwindung* (surpassing) but must be embraced in the nuanced conjunction of two other terms: *Andenken* (remembrance) and *Verwindung* (distortion, twisting, incorporation). The two characteristic aspects of *Verwindung* are combined in the dyadic expression acceptance-distortion.
3. "Well, well, let's get on with it..." are the closing words in Sartre's 1944 play *Huis Clos* (*Eyes Closed*) in the 1989 English translation by Stuart Gilbert (the title of play was translated by Gilbert as *No Exit*). The final line of the play, uttered by the main character, Garcin, expresses the kind of resignation that cannot but characterize any effort directed at understanding or altering the human condition, including this book.

Bibliography

Aarseth, Espen J. 1997. *Cybertext – Perspectives on Ergodic Literature*. Baltimore: The John Hopkins University Press.

Bell, Daniel. 1974. *The Coming of Post-Industrial Society: A Venture in Social Forecasting*. London (UK): Heinemann.

Benjamin, Walter. 2008 (1936). *The Work of Art in the Age of Its Technological Reproducibility, and Other Writings on Media*. Cambridge (MA): Belknap Press of Harvard University Press.

Berger, P. L., Luckman, T. 1988. *The Social Construction of Reality*, New York (NY): Anchot.

Bijker, W., Law, J. 1992. *Shaping Technology/Building Society: Studies in Sociotechnical Change*. Cambridge (MA): The MIT press.

Bissell, Christopher Charles. 2011. "Hermann Schmidt and German 'proto-cybernetics'". Published in *Information, Communication & Society*, 14(1), 156–171. Available online at http://oro.open.ac.uk/26141/2/2B5B59E1.pdf (accessed on February 9, 2013).

Black, Max. 1962. *Models and Metaphors: Studies in Language and Philosophy*, Ithaca (NY): Cornell University Press.

Boal, Augusto. 1998. *The Theatre of the Oppressed*. New York (NY): TGC.

Boccignone, Martino E. 2009. "The Duty of Personal Identity: Authenticity and Irony". Proceedings of the fourth International Plessner Conference. Available online at http://independent.academia.edu/MartinoEnricoBoccignone/Papers/166390/The_Duty_of_Personal_Identity_Authenticity_and_Irony (accessed on February 9, 2013).

Bogost, Ian. 2005. "Frame and Metaphor in Political Games". Proceedings of the DiGRA 2005 Conference: Changing Views – Worlds in Play. Available online at http://www.digra.org/dl/db/06276.36533.pdf (accessed on February 9, 2013).

Bogost, Ian. 2007. *Persuasive Games: The Expressive Power of Videogames*. Cambridge (MA): The MIT Press.

Bogost, Ian. 2011. *How to Do Things with Videogames*. Minneapolis (MN): University of Minnesota Press.

Bogost, Ian. 2012. *Alien Phenomenology, or What It's Like to Be a Thing*. Minneapolis (MN): University of Minnesota Press.

Bolter, J. D. 2001. *Writing Space: Computers, Hypertext, and the Remediation of Print, Second Edition*. Mahwah (NJ): Lawrence Erlbaum Associates.

Bolter, J. D. 2003. "Theory and Practice in Media Studies". In Liestøl, G., Morrison, A., Rasmussen, T. (edited by). *Digital Media Revisited*. Cambridge (MA): The MIT Press. 15–34.

Bolter, J. D., Grusin, R. 2000 (1998). *Remediation: Understanding New Media*. Cambridge (MA): The MIT Press.

Borges, Jorge Luis. 1994 (1944). *Ficciones*. New York (NY): Grove Press.

Borges, Jorge Luis. 2001. *The Total Library*. London (UK): Penguin Books Ltd.

Borges, Jorge Luis. 2004. *The Aleph and Other Stories*. London (UK). Penguin Books Ltd.

Bostrom, Nick. 2003. "Are You Living in a Computer Simulation?" In *Philosophical Quarterly*, Vol. 53, N. 211, 243–255. Also available online at http://www.simulation-argument.com/simulation.html (accessed on February 9, 2013).

Boswall, Jeffery. 1977. "Tool-using by birds and related behaviour". In *Avicultural Magazine*, N. 83. 88–97. Available online at http://ia600507.us.archive.org/8/items/ToolUsingBirds/ToolUsing.pdf (accessed on February 9, 2013).

Brindle, John. 2013. "The McDonalds sim and September 12: what does it mean for a videogame to be political?" Available online at http://www.newstatesman.com/culture/2013/02/mcdonalds-sim-and-september-12-what-does-it-mean-videogame-be-political (accessed on February 20, 2013)

Burke, Kenneth. 1941. "Four Master Tropes". In *The Kenyon Review* 3, 1941, Vol. 4, 421–438.

Cappuccio, Massimiliano. 2005. *Alan Turing: l"uomo, la macchina, l"enigma – Per una genealogia dell'incomputabile*. Milano (Italy): Edizioni Albo Versorio.

Carman, Taylor. 1999. "The Body in Husserl and Merleau-Ponty". Published in *PHILOSOPHICAL TOPICS*, Vol. 27, N. 2, Fall 1999, 205–226. Available online at http://ist-socrates.berkeley.edu/~hdreyfus/188_s05/pdf/Carman_Body.pdf (accessed on February 9, 2013).

Carr, Nicholas. 2010. *The Shallows: What the Internet is Doing to our Brains*. New York (NY): W. W. Norton & Company Inc.

Chandler, Daniel. 2007. *Semiotics. The Basics*. 2nd ed. New York (NY): Routledge.

Churchland, Paul. 1988. *Matter and Consciousness, Revised Edition*. Cambridge (MA): The MIT Press.

Clark, Taylor. 2012. "The Most Dangerous Gamer", in *The Atlantic*, May 2012 issue. Available online at http://www.theatlantic.com/magazine/archive/2012/05/the-most-dangerous-gamer/8928/ (accessed on February 9, 2013).

Compagno, Dario S. 2008. I am Link's Transcendental Will: Freedom from Hyrule to Earth. In L. Cuddy (edited by). The Legend of Zelda and Philosophy. Chicago (IL): Open Court.

Coolen, Maarten. 1992. *De machine voorbij. Over het zelfbegrip van de mens in het tijdperk van de informatietechniek*. Amsterdam (The Netherlands): Boom.

Coolen, Maarten. 2013. "Bodily Experience and Experiencing One's Body". In De Mul, Jos (edited by). *ARTIFICIAL BY NATURE: Plessner's Philosophical Anthropology: Perspectives and Prospects*. In the process of being published in Amsterdam (The Netherlands): University of Amsterdam Press. 65–74.

Costa, Mario. 2007. *La disumanizzazione tecnologica – il destino dell'arte nell'epoca delle nuove tecnologie*. Milano (Italy): Costa & Nolan.

DeKoven, Bernard. 2002. *The Well-Played Game. a Playful Path to Wholeness*. Lincoln (NE): Writers Club Press.

De Mul, Jos. 1999. "The Informatization of the worldview". In *Information, Communication & Society*. 1999. Vol. 2, N. 1, 69–94. Available online at http://www.demul.nl/nl/publicaties/publicaties-per-categorie/artikelen/item/1584-the-informatization-of-the-worldview (accessed on March 7, 2013).

De Mul, Jos. 2002. *Een introductie*. In De Mul, Jos (edited by). *Filosofie in Cyberspace: Reflectie op de informatie- en communicatietechnologie*. Kampen (The Netherlands): Klement. 9–44.

De Mul, Jos. 2004. *The Tragedy of Finitude: Dilthey's Hermeneutics of Life*. New Haven (CT): Yale University Press.

De Mul, Jos. 2007. "Wittgenstein 2.0: Philosophical reading and writing after the mediatic turn". In *Wittgenstein and the Philosophy of Information* – proceedings of the 30th international Ludwig Wittgenstein Symposium, Volume I, edited by Pichler, A. and Hrachovec, H. Kirchberg am Wechsel (Austria). 153–179.

De Mul, Jos. 2010 (2002). *Cyberspace Odyssey: Towards a Virtual Ontology and Anthropology*. Cambridge (MA): Cambridge Scholars Publishing.

De Mul, Jos (edited by). 2013. *ARTIFICIAL bY NATURE: Plessner's Philosophical Anthropology: Perspectives and Prospects*. In the process of being published in Amsterdam (The Netherlands): University of Amsterdam Press.

Derrida, Jacques. 1981. "Plato's Pharmacy". Available in Derrida, Jacques. 1981. *Dissemination*. Trans. Barbara Johnson. London (UK): Athlone Press. 61–171.

Descartes, René. 1637. *Discourse on the Method of Rightly Conducting the Reason and Seeking the Truth in Science*. Available online at http://www.gutenberg.org/files/59/59-h/59-h.htm (accessed on March 7, 2013).

Descartes, René. 1641. *Meditations on First Philosophy*. Available online at http://www.filepedia.org/files/Descartes"%20Meditations%20on%20First%20Philosophy.pdf (accessed on March 7, 2013).

Dick, Philip K. 1978. "How to Build a Universe That Doesn't Fall Apart Two Days Later". Available online at http://deoxy.org/pkd_how2build.htm (accessed on February 9, 2013).

Diels, H. and Kranz, W. 1903. *Die Fragmente der Vorsokratiker*. Frankfurt Am Main (Germany): Weidmannsche buchhandlung.

Dormans, Joris. 2011. "Beyond Iconic Simulation". In *Simulation & Gaming*. 2011. Vol. 42, N. 5, 610–631.

Dreyfus, H., Spinoza, C. 2006. "Further Reflections on Heidegger, Technology and the Everyday". In Kompridis, Nicholas (edited by). *Philosophical Romanticism*. New York (NY): Routledge. 265–281.

Dreyfus, Hubert L. 1979. *What Computers Can't Do*. New York (NY): Harper & Row Publishers Inc.

Dreyfus, Hubert L. 1991. *Being-in-the-World – A commentary on Heidegger's Being and Time, Division I*. Cambridge (MA): The MIT Press.

Dreyfus, Hubert L. 2000. *Telepistemology: Descartes's Last Stand*. In Goldberg, Ken (edited by). *The Robot in the Garden: Telerobotics and Telepistemology in the Age of the Internet*. Cambridge (MA): The MIT Press. 48–63.

Dunne, A. & Raby, F. 2013. *Speculative Everything: Design, Fiction, and Social Dreaming*. Cambridge, MA: The MIT Press.

Emden, Christian J. 2005. *Nietzsche on Language, Consciousness, and the Body*. Champaign (IL): University of Illinois Press.

Fedier, Francois. 2001. *L"Arte. Aristotele, Cézanne, Matisse: Il pensiero in pittura*. Milano (Italy): Christian Marinotti edizioni.

Fischer, Joachim. 2013. Philosophical Anthropology – A third way between Darwinism and Foucaultism. In De Mul, Jos (edited by). *ARTIFICIAL bY NATURE: Plessner's Philosophical Anthropology: Perspectives and Prospects*. In the process of being published in Amsterdam (The Netherlands): University of Amsterdam Press. 19–26.

Flanagan, Mary. 2009. *Critical Play. Radical Game Design*. Cambridge (MA): The MIT Press.

Flusser, Vilém. 1992. "Das Ende der Tyrannei". Arch + Zeitschrift für Architektur und Stadtebau (March 1992), 20–25.
Føllesdal, Dagfinn. 1979. "Husserl and Heidegger on the Role of Actions in the Constitution of the World". In Saarinen, Esa et al. (edited by). *Essays in Honour of Jakko Hintikka*. Dodrecht (The Netherlands): Reidel. 365–378.
Forster, Edward Morgan. 1985 (1909). *The Machine Stops* in *The New Collected Short Stories*. London (UK): Sidgwick & Jackson.
Foucault, Michel. 1982. "The Subject and Power". in *Michel Foucault: Beyond Structuralism and Hermeneutics*. 1982. 2nd edition. Ed. Dreyfus, H. L. & Rabinow, P. Chicago (IL): University of Chicago Press.
Foucault, Michel. 1988. *Technologies of the Self: A Seminar with Michel Foucault*. Edited by Martin L. H., Gutman, H. and Hutton P. H. London (UK): Tavistock Publications.
Foucault, Michel. 1994 (1966). *The Order of Things – an Archaeology of the Human Sciences*. New York (NY): Vintage Books.
Frasca, Gonzalo. 2003. "Simulation versus Narrative: Introduction to Ludology". The essay is available online at http://ludology.org/articles/VGT_final.pdf (accessed on February 9, 2013).
Galloway, Alexander R. 2006. *Gaming: Essays on Algorithmic Culture*. Minneapolis (MN): University of Minnesota Press.
Gandy, Robin O. 1988. *The Confluence of Ideas in 1936*. In Herken, Rolf (edited by). *The Universal Turing Machine – A Half-Century Survey*. New York (NY): Oxford University Press. 55–111.
Garrett, Brian. 2006. *What Is This Thing Called Metaphysics?* New York (NY): Routledge.
Gay, Peter. 1966. *The Enlightenment: An Interpretation*. New York (NY): W. W. Norton & Company.
Geymonat, Ludovico. 1971. *Storia del Pensiero Filosofico e Scientifico* (Vol. III). Milano (Italy): Aldo Garzanti Editore.
Geymonat, Ludovico. 1972. *Storia del Pensiero Filosofico e Scientifico* (Vol. VI). Milano (Italy): Aldo Garzanti Editore.
Giner-Sorolla, Roger, 1996. "Crimes against Mimesis". Originally posted to Usenet in April 1996 and published online by Stephen van Egmond in January 1998. Available online at http://web.archive.org/web/20050619081931/http://www.geocities.com/aetus_kane/writing/cam.html (accessed on February 9, 2013).
Gold, Matthew K. (edited by). 2012. *Debates in the Digital Humanities*. Minneapolis (MN): The University of Minnesota Press.
Goldberg, Ken (edited by). 2000. *The Robot in the Garden: Telerobotics and Telepistemology in the Age of the Internet*. Cambridge (MA): The MIT Press.
Gualeni, Stefano. 2012a. "La Voce Del Sonnambulismo Contemporaneo". Article published in the critical section of the first issue of *G|A|M|E – The Italian Journal of Game Studies*, May the 29th 2012. Available online at http://www.game-journal.it/italiano-la-voce-del-sonnambulismo-contemporaneo/ (accessed on February 9, 2013).
Gualeni, Stefano. 2012b. 'super Mafio Bros". Column published on Monday, February the 20th 2012 on *Control* Magazine. Available online at http://www.control-online.nl/gamesindustrie/2012/02/20/column-brieven-uit-deanderwereld-deel-5-super-mafio-bros (accessed on February 9, 2013).

Gualeni, Stefano. 2014a. "Augmented Ontologies; or, How to Philosophize with a Digital Hammer". Journal article published in Springer's *Philosophy of Technology*. 2013. Edited by Luciano Floridi, Vol. 27, N. 2, 177–199. Springer Netherlands. ISSN 2210–5433, Philos. Technol., DOI 10.1007/s13347–013–0123-x

Gualeni, Stefano. 2014b. "FREER THAN WE THINK: Game Design as a Liberation Practice". Paper presented at the 2014 Philosophy of Computer Games conference, held at the Bilgi University of Istanbul, Turkey, November 13–16, 2014.

Gualeni, Stefano. 2015. "Playing with Puzzling Philosophical Problems". In Zagalo, N. and Branco, P. (eds.) 2015. Creativity in the Digital Age. Part of the Springer Series on Cultural Computing. XIV. London (UK): Springer.

Hansen, Mark B.N. 2003. *New Philosophy for New Media*. Cambridge, MA: The MIT Press.

Haraway, Donna. 1991. "A Cyborg Manifesto: Science, Technology, and Socialist-Feminism in the Late Twentieth Century", in Haraway, Donna. 1991. *Simians, Cyborgs and Women: The Reinvention of Nature*. New York (NY): Routledge. 149–181.

Heidegger, Martin. 1936. "The Origin of the Work of Art". In Heidegger, Martin. 2008. *Basic Writings*. New York (NY): Harper Perennial – Modern Thought. 143–212.

Heidegger, Martin. 1962 (1927). *Being and Time*. Trans. Macquarrie, J. and Robinson, E. San Francisco (CA): Harper & Row Publishers Inc.

Heidegger, Martin. 1966. "The Memorial Address". In *Discourse on Thinking*. Trans. Anderson, J. M. and Freund, E. H. New York (NY): Harper & Row Publishers Inc.

Heidegger, Martin. 1972 (1962). *On Time and Being*. New York (NY): Harper & Row Publishers Inc.

Heidegger, Martin. 1973 (1969). *The End of Philosophy*. New York (NY): Harper & Row Publishers Inc.

Heidegger, Martin. 1977. *Gesamtausgabe*. Vol. 5. In De Mul, Jos. 2004. *The Tragedy of Finitude: Dilthey's Hermeneutics of Life*. New Haven (CT): Yale University Press. 310–311.

Heidegger, Martin. 1982. *The Question Concerning Technology and Other Essays*. New York (NY): Harper & Row Publishers Inc.

Heidegger, Martin. 1998. *Pathmarks*. Cambridge (MA): Cambridge University Press.

Heidegger, Martin. 2000 (1953). *Introduction to Metaphysics*. Trans. Richard Polt and Gregory Fried. New Haven: Yale University Press.

Heidegger, Martin. 2002 (1969). *Identity and Difference*. Chicago (IL): University of Chicago Press.

Heidegger, Martin. 2008. *Basic Writings*. New York (NY): Harper Perennial – Modern Thought.

Heidegger, Martin. 2011 (1924). *The Concept of Time*. Trans. Ingo Farin. London (UK): Continuum.

Heim, Michael. 1993. *The Metaphysics of Virtual Reality*. Oxford (UK): Oxford University Press.

Hodges, Andrew. 1992. *Alan Turing: The Enigma*. London (UK): Vintage.

Hofstadter, A., Kuhns, R. (edited by). 1976. *Philosophies of Art and Beauty*. London (UK): University of Chicago Press.

Hughes, Thomas P. 1994. *Technological Momentum*. In Smith, M. R., Marx, L. (edited by). *Does Technology Drive History? The Dilemma of Technological Determinism*. Cambridge (MA): The MIT Press. 101–113.

Huhtamo, Erkki. 1995. "Encapsulated bodies in motion: Simulation and the quest for Total Immersion". In Penny, Simon (edited by). *Critical Issues in Electronic Media*. Albany (NY): State University of New York Press. 159–186.

Huizinga, Johan. 1955 (1938). *Homo ludens; a study of the play-element in culture*. Boston (MA): Beacon Press.

Hume, David. 1738. *A Treatise of Human Nature*. Available online at http://www.gutenberg.org/files/4705/4705-h/4705-h.htm (accessed on May 28, 2014).

Hume, David. 1748. *An Enquiry Concerning Human Understanding*. Available online at http://www.gutenberg.org/files/9662/9662-h/9662-h.htm (accessed on May 28, 2014).

Hume, David. 1771. *Dialogues Concerning Natural Religion*. Available online at http://www.anselm.edu/homepage/dbanach/dnr.htm (accessed on May 28, 2014).

Ihde, Don. 1986. *Experimental Phenomenology: An Introduction*. Albany (NY): State University of New York Press.

Ihde, Don. 1990. *Technology and the Lifeworld*. The Indiana series in the Philosophy of Technology. Bloomington (IN): Indiana University Press.

Ihde, Don. 1993. *Postphenomenology: Essays in the Postmodern Context*. Evanston (IL): Northwestern University Press.

IJsselstijn, W., Harper, B. 2001. "Virtually There? A vision on Presence Research". In *PRESENCE* – IST 2000–31014, EC Public Deliverable (D2), December 2001. Also available online at ftp://ftp.cordis.europa.eu/pub/ist/docs/fet/fetpr-4.pdf (accessed on February 9, 2013).

Iriki, A., Tanaka, M., Iwamura, Y. 1996. "Coding of modified body schemas during tool use by macaque postcentral neurons". In Neuroreport, N. 7, 2325–2330.

Irwin, William (edited by). 2005 (2002). *The Matrix and Philosophy*. Chicago (IL): Open Court Publishing Company.

Jaspers, Karl T. 1951 (1931). *Man in the Modern Age*. Trans E. Paul and C. Paul. London (UK): Routledge & Kegan Paul.

Jauss, Hans Robert. 2007 (1982). *Toward an Aesthetic of Reception*. Minneapolis (MN): University of Minnesota Press.

Journet, Debra. 2005. "Metaphor, Ambiguity, and Motive in Evolutionary Biology". W.D. Hamilton and the "Gene's Point of View" in Written Communication, Vol. 22, N. 4, 379–420.

Journet, Debra. 2010. "The resources of Ambiguity. Context, Narrative and Metaphor in Richard Dawkins's "The Selfish Gene". In *Journal of Business and Technical Communication*, 2010, Vol. 24, N. 1, 29–59.

Juul, Jesper. 2003. "The Game, the Player, the World: Looking for a Heart of Gameness". In Copier, M. and Raessens, J. (edited by). *Level Up: Digital Games Research Conference Proceedings*. Utrecht (The Netherlands): Utrecht University Press. 30–45.

Juul, Jesper. 2004. "Introduction to Game Time / Time to Play". In Wardrip-Fruin, N., Harrigan, P. (edited by). *First Person: New Media as Story, Performance, and Game*. Cambridge (MA): The MIT Press. 131–142. Available online at http://www.jesperjuul.net/text/timetoplay (accessed on February 9, 2013).

Juul, Jesper. 2005. *Half-Real: Video Games between Real Rules and Fictional Worlds*. Cambridge (MA): The MIT Press.

Juul, Jesper. 2009. *A Casual Revolution: Reinventing Video Games and Their Players*. Cambridge (MA): The MIT Press.

Kant, Immanuel. 1783. *Prolegomena to Any Future Metaphysics That Will Be Able to Present Itself as a Science*. Available online at http://www.webexhibits.org/causesofcolor/ref/Kant.html (accessed on September 8, 2013).

Kant, Immanuel. 2000 (1998). *Critique of Pure Reason*. Trans. Guyer, P. and Wood, A. W. Cambridge (MA): Cambridge University Press.

Kapp, Ernst. 1877. *Grundlinien einer Philosophie der Technik*. Braunschweig (Germany): George Westermann ed. Available online at http://vlp.mpiwg-berlin.mpg.de/library/data/lit39532 (accessed on September15, 2013).

Kline, Stephen J. 2003. *What is Technology?* In Robert C. Scharff and Val Dusek (edited by). *Philosophy of technology: The technological condition: An anthology*. Malden (MA): Blackwell publishers. 210–213.

Kövecses, Zoltán. 2010. *Metaphor: A Practical Introduction*. 2nd ed. New York (NY): Oxford University Press.

Krocker, Arthur. 2004. *The Will to Technology and the Culture of Nihilism. Heidegger, Nietzsche and Marx*. Toronto (Canada): University of Toronto Press.

Kurzweil, Raymond. 2003. "Kurzweil's Law (aka "the law of accelerating returns")". Available online at http://www.kurzweilai.net/articles/art0610.html?printable=1 (accessed on December 25, 2009).

Lahusen, Susanne. 1986. "Oskar Schlemmer: Mechanical Ballets?" In *Dance Research*, Autumn 1986, Vol. 4, Issue 2, 65.

Lakoff, G., Johnson, M. 2003 (1980). *Metaphors We Live By*. Chicago (IL): University of Chicago Press.

Lévy, Pierre. 1998. *Qu'est que le Virtuel?* Paris: La Découverte.

Lévy, Pierre. 2001 (1997). *Cyberculture*. Minneapolis (MN): University of Minnesota Press.

Loeffler, C. E., Anderson, T. 1994. *The Virtual Reality Casebook*. New York (NY): Van Nostrand Reinhold.

Mackenzie, D. A., Wajcman, J. 1999. *The Social Shaping of Technology*. 2nd ed. Buckingham (UK); Philadelphia (PA): Open University Press.

Magnus, Bernd. 1970. *Heidegger's Metahistory of Philosophy: Amor Fati, Being and Truth*. Den Haag (The Netherlands): Martinus Nijhoff.

Mancini, G., Marzocchi, S., Picinali, G. 1993. *Corso di Filosofia (Storia)*. Milano (Italy): Bompiani.

Manovich, Lev. 2001 (2000). *The Language of New Media*. Cambridge (MA): The MIT Press.

Marcuse, Herbert. 1991 (1964). *One-Dimensional Man*. Boston: Beacon Press.

Margreiter, Reinhard. 2003. *Medien/Philosophie ein Kippbild*. In *Medienphilosophie: Beiträge zur Klärung eines Begriffs*, edited by Münker, S., Roesler, A. and Sandbothe, M. Frankfurt am Main: Fischer Taschenbuch Verlag.

Masterman, Margaret. 1962. "The Intellect's New Eye". In *Times Literary Supplement*, N. 284 (April the 17th 1962). Available as a reprint in *Freeing the Mind: Articles and Letters from the Times Literary Supplement during March-June 1962*. London (UK): Times. 38–44.

McCarthy, Willard. 2005. *Humanities Computing*. New York (NY): Palgrave.

McCarthy, Willard. 2012. "A Telescope for the Mind?" In Gold, Matthew K. (edited by). *Debates in the Digital Humanities*. Minneapolis (MN): The University of Minnesota Press. 113–123. Available online at http://www.mccarty.org.uk/essays/McCarty,%20Telescope.pdf (accessed on September 11, 2013).

McLuhan, Marshall. 1994 (1964). *Understanding Media: The Extensions of Man*. Cambridge (MA): The MIT Press.

McMahan, Alison. 2003. *Immersion, Engagement, and Presence: A Method for Analyzing 3-D Video Games*. In Wolf, M. J. P. & Perron, B. 2003. *The Video Game Theory Reader*. London (UK): Routledge. 67–86.

Mead, G. H. 2001. *Play, School and Society*. New York (NY): Peter Lang.

Meillassoux, Quentin. 2011 (2008). *After Finitude*. New York (NY): Continuum International Publishing Group.

Merleau-Ponty, Maurice 1962 (1945). *Phenomenology of Perception*. London (UK): Routledge.

Minsky, Marvin. 1980. "Telepresence. Omni Magazine". Available online at http://web.media.mit.edu/~minsky/papers/Telepresence.html (accessed on February 9, 2013).

Möring, Sebastian. 2012. "Tackling the Metaphor-Simulation Dilemma". Proceedings of DiGRA Nordic 2012 Conference: Local and Global – Games in Culture and Society. Available online at http://www.digra.org/dl/db/12168.04280.pdf (accessed on February 9, 2013).

Mosca, Ivan. 2012. "+10! Gamification and deGamification". In *G|A|M|E – The Italian Journal of Game Studies*, May the 29th 2012. Available online at http://www.gamejournal.it/plus10_gamification-and-degamification/ (accessed on February 9, 2013).

Mosca, Ivan. 2013. "From Fiction to Reality and Back: Ontology of Ludic Simulations", in the *International Journal of Gaming and Computer-Mediated Simulations*, issue 5(1), January-March 2013, 13–31.

Murray, Janet H. 2003. *Inventing the Medium*. In Wardrip-Fruin, N., Montfort N. (edited by). *The New Media Reader*. Cambridge (MA): The MIT Press. 3–5.

Musil, Robert. 1996 (1930–1943). *The Man Without Qualities*. Vol. I. New York (NY): Vintage International.

Nagel, Thomas. 1974. "What Is it Like to Be a Bat?" In *The Philosophical Review*, 1974, Vol. 83, 435–450. Ithaca (NY): Cornell University Press. Available online at http://organizations.utep.edu/Portals/1475/nagel_bat.pdf (accessed on February 9, 2013).

Nagel, Thomas. 1989 (1986). *The View from Nowhere*. New York (NY): OUP USA.

Nietzsche, Friedrich. 2013 (1872). *The Birth of Tragedy from the Spirit of Music*. Trans. Ian Johnston. San Bernardino (CA): Maestro Reprints.

Nitsche, Michael. 2009. *Video Game Spaces: Image, Play, and Structure in 3D Worlds*. Cambridge (MA): The MIT Press.

Nye, David E. 2006. *Technology Matters: Questions to Live with*. Cambridge (MA): The MIT Press.

Nyíri, J. Kristóf. 1993. "Thinking With a Word Processor". Lecture given in Kirchberg (Austria) at the 1993 Wittgenstein symposium organized by Austrian Ludwig Wittgenstein Society. Available online at http://www.hunfi.hu/nyiri/KRB93_TLK.htm#N_1_ (accessed on February 9, 2013).

Nyíri, J. Kristóf. 1996. "Wittgenstein as a Philosopher of Secondary Orality". In *Grazer Philosophische Studien – International Journal for Analytic Philosophy*, 1996/97, Vol. 52, 45–57. Available online at http://hunfi.hu/nyiri/gps97.htm (accessed on February 9, 2013).

Ong, Walter J. 1982. *Orality and Literacy: The Technologizing of the Word*. London (UK): Methuen.

O'sullivan, Edmund. 2003. "Bringing a perspective of transformative learning to globalized consumption". *International Journal of Consumer Studies*, 27 (4), 326–330.

Panza, Pierluigi. 2002. *Estetica, tempo e progetto nell'età delle comunicazioni*. Milano (Italy): Guerini Studio Editore.

Parker, Felan. 2011. "In the Domain of Optional Rules: Foucault's Aesthetic Self-Fashioning and Expansive Gameplay". Paper presented at the 2011 Games and Philosophy Conference, annual conference held at the Panteion University of Athens, Greece, April 6–9, 2011.

Pedercini, Paolo. 2014. "Making Games in a Fucked-up World". Blogpost available online at http://www.molleindustria.org/blog/making-games-in-a-fucked-up-world-games-for-change-2014 (accessed on October 15, 2014)

Pimentel, K., and Teixeira, K. 1993. *Virtual Reality: Through the New Looking Glass*. New York (NY): Windcrest Books.

Pirsig, Robert M. 1984. *Zen and the Art of Motorcycle Maintenance: An Inquiry into Values*. New York (NY): Bantam Books.

Plato, 1981. *Meno* (Translated, with Introduction and Notes, by G.M.A. Grube). Indianapolis (IN): Hackett Publishing Company.

Plato, 1995. *Phaedrus* (Translated, with Introduction and Notes, by Alexander Nehamans and Paul Woodruff). Indianapolis (IN): Hackett Publishing Company.

Plessner, Helmuth. 1975. *Die Stufen des Organischen und der Mensch*. Einleidtung in die philosophische Anthropologie. Vol. IV. *Gesammelte Schriften*. Frankfurt a/M: Suhrkamp.

Plessner, Helmuth. 1980–1985. *Gesammelte Schriften*. Frankfurt am Main (Germany): Suhrkamp.

Plessner, Helmuth. 1998 (1959). *Die verspätete Nation. Über die politische Verführbarkeit bürgerlichen Geistes*. Frankfurt am Main (Germany). 9–31.

Plessner, Helmuth. 2006 (1928). *I gradi dell'organico e l"uomo. Introduzione all'antropologia filosofica*. Torino (Italy): Bollati Boringhieri.

Postman, Neil. 1992. *Technopoly: The Surrender of Culture to Technology*. New York (NY): Vintage.

Postman, Neil. 2005 (1986). *Amusing Ourselves to Death: Public Discourse in the Age of Show Business*. London (UK): Penguin Books Ltd.

Queneau, Raymond. (1961). *Cent Mille Milliards de Poèmes*. Paris (France): Éditions Gallimard.

Ramsay, S. & Rockwell, G. 2012. "Developing Things: Notes towards an Epistemology of Building in the Digital Humanities". In Gold, Matthew K. (edited by). *Debates in the Digital Humanities*. Minneapolis (MN): The University of Minnesota Press. 75–84.

Rayner, Timothy. 2007. *Foucault's Heidegger: Philosophy and Transformative Experience*. London (UK): Bloomsbury Academic.

Rasini, Vallori. 2010. "Perche" un"antropologia filosofica: le motivazioni di Helmut Plessner". In *Etica & Poilitica / Ethics & Politics*, Vol. XII, N. 2, 164–177. Available online at http://www2.units.it/etica/2010_2/RASINI.pdf (accessed on February 9, 2013).

Richard Villa, Dana. 1996. *Arendt and Heidegger: The Fate of the Political*. Princeton (NJ): Princeton University Press.

Richarson, William J. 1967. *HEIDEGGER: Through Phenomenology to Thought*, Den Haag (The Netherlands): Martinus Nijhoff.

Ricoeur, Paul. 2008 (1975). *The Rule of Metaphor*. Toronto (Canada): Toronto UP.

Rivas, Virgilio Aquino. 2007. "The 'Turn' to Time and the Miscarriage of Being". In *Kritike*, Vol. 1, N. 2, 65–81. Available online at http://www.kritike.org/journal/issue_2/rivas_december2007.pdf (accessed on February 9, 2013).

Rorty, Richard McKay. 1967. *The Linguistic Turn*. Chicago (IL): University of Chicago Press.

Russell, Stuart J., Norvig, P. 2003. *Artificial Intelligence: A Modern Approach* (2nd ed.), Upper Saddle River (NJ): Prentice Hall.

Safranski, Rüdiger. 2002 (1998). "Martin Heidegger: between Good and Evil". Trans. Ewald Osers. Cambridge (MA): Harvard University Press.

Salen, K., Zimmerman, E. 2003. *Rules of Play: Game Design Fundamentals*. Cambridge (MA): The MIT Press.

Sartre, Jean Paul. 1989 (1944). *No Exit and Three Other Plays*. Trans. Stuart Gilbert. New York (NY): Vintage books.

Sartre, Jean Paul. 2010 (1943). *Being and Nothingness*. Abingdon (UK): Routledge Classics.

Schmidt, Hermann. 1954. "Die Entwicklung der Technik als Phase der Wandlung des Menschen". In *Zeitschrift des Vereins Deutsche Ingenieure*. 1954. 118–122.

Shanker, Stuart G. 1987. "Wittgenstein versus Turing on the nature of Church's thesis". In Notre Dame J. (edited by). *Formal Logic*, Volume 28, N. 4, 615–649.

Sheehan, Thomas. 1984. "Heidegger's Philosophy of Mind". In Floistad, Guttorm (edited by). *Contemporary Philosophy: A New Survey*. The Hague (The Netherlands): Nijnhoff. 287–318.

Sicart, Miguel. 2008. "Defining Game Mechanics". In *Game Studies – the International Journal of Game Studies*, volume 8, issue 2, December 208 ISSN: 1604–7982. Available online at http://gamestudies. org/0802/articles/sicart

Sicart, Miguel. 2011. "Against Procedurality". In *Game Studies – the International Journal of Game Studies*, volume 11, issue 3, December 2011 ISSN: 1604–7982. (Available online at: http://gamestudies.org/1103/articles/sicart_ap)

Smith, Merritr R., Marx, Leo. 1994. *Does Technology Drive History?* Cambridge (MA): The MIT Press. (accessed on May 28, 2014).

Smith, Ed. (2012). *A Personal Journey: Jenova Chen's Goals for Games*. Article featured on Gamasutra.com on May the 18th 2012. Available online at http://www.gamasutra.com/view/feature/170547/a_personal_journey_jenova_chens_.php

Stern, David G. 1994. "Wittgenstein Papers as Text and Hypertext: Cambridge, Bergen and Beyond". In Solum Forlag (edited by). *WITTGENSTEIN and Norway*. Larvik (Norway): Østlands-Postens Boktrykkeri. 251–269. (accessed on May 28, 2014).

Stern, David G. 1996. "The availability of Wittgenstein's philosophy". In Sluga H. D. and Stern D. G. (edited by). *The Cambridge companion to Wittgenstein*. New York (NY): Cambridge University Press. 442–476.

Stillings, Neill et Al. 1995. *Cognitive Science: An Introduction*. Cambridge (MA): The MIT Press.

Stone, Abraham. 2005. "Heidegger and Carnap on the Overcoming of Metaphysics". In VV. AA., 2006. *Martin Heidegger*. Ashgate: Ed. Stephen Mulhall. Available online at http://people.ucsc.edu/~abestone/papers/uberwindung.pdf (accessed on February 9, 2013).

Ströhl, Andreas (edited by). 2004. *Vilém Flusser – Writings*. Minneapolis (MN): University of Minnesota Press.

Tabachnick, David Edward. 2006. "The Tragic Double Bind of Heidegger's Techné". In *PhaenEx 1*, N. 2, (fall/winter 2006), 94–112. Available online at http://www.phaenex.uwindsor.ca/ojs/leddy/index.php/phaenex/article/viewFile/226/233 (accessed on February 9, 2013).

Thompson, Iain Donald. 2005. *Heidegger on Ontotheology: Technology and the Politics of Education*. Cambridge (MA): Cambridge University Press.

Turing, Alan. 1936. "On Computable Numbers, with an Application to the Entscheidungsproblem". Available online at http://www.thocp.net/biographies/papers/turing_oncomputablenumbers_1936.pdf (accessed on February 9, 2013).

Turkle, Sherry. 2007 (2005). *Life on the Screen*. New York, NY: Touchstone.

Umiltà, M.A., Escola, L., Intskirveli, I., Grammont, F., Rochat, M., Caruana, F., Jezzini, A., Gallese, V., Rizzolatti, G., 2008, "When pliers become fingers in the monkey motor system". In *PNAS*, February 12, 2008, Vol. 105, N. 6, 2209–2213. Available online at http://www.pnas.org/content/105/6/2209.full (accessed on February 9, 2013).

Valéry, Paul. 1939. "Poésie et Pensée Abstraite". In *Oeuvres*, 1957, Paris: Gallimard, Vol. I, 1324–1356.

Valve Corporation. 2007. *Portal*. Windows.

Van Binsbergen, Wim. 1997. *Virtuality as a Key Concept in the Study of Globalisation*. Den Haag (The Netherlands): WOTRO.

Van den Berg, Bibi. 2009. *The Situated Self: Identity in a World of Ambient Intelligence*. Rotterdam (The Netherlands): Erasmus University.

Vattimo, Gianni. 1963. *Essere, storia e linguaggio in Heidegger*. Torino (Italy): Edizioni di "Filosofia"

Vattimo, Gianni. 1991 (1985). *The End of Modernity*. Baltimore (MD): The John Hopkins University Press.

Vattimo, Gianni. 2004 (2003). *Nihilism and Emancipation*. New York (NY): Columbia University Press.

Vattimo, Gianni. 2005 (1971). *Introduzione ad Heidegger*. Bari (Italy): Editori Laterza. (Tenth volume of the Italian collection *I Filosofi* – The Philosophers).

Verbeek, Peter-Paul. 2005 (2000). *What Things Do*. Trans. Robert P. Crease. University Park (PA): The Pennsylvania University Press.

Verbeek, Peter-Paul. 2011. *Moralizing Technology: Understanding and Designing the Morality of Things*. Chicago (IL): The University of Chicago Press.

Verbeek, Peter-Paul. 2013. *Plessner and Technology: Philosophical Anthropology Meets the Posthuman*. In De Mul, Jos (edited by). *ARTIFICIAL BY NATURE: Plessner's Philosophical Anthropology: Perspectives and Prospects*. In the process of being published in Amsterdam (The Netherlands). University of Amsterdam Press. 231–237.

Volpi, Franco. 1998. "Heidegger e i Greci". In Pinelli, E. (edited by). *I filosofi antichi nel pensiero del Novecento*. Roma (Italy): Ministero della Pubblica Istruzione. 87–104.

Wardrip-Fruin, N., Harrigan, P. (edited by). 2004. *First Person – New Media as Story, Performance and Game*. Cambridge (MA): The MIT Press.

Wardrip-Fruin, N., Harrigan, P. (edited by). 2007. *Second Person – Role-Playing and Story in Games and Playable Media*. Cambridge (MA): The MIT Press.

Wardrip-Fruin, N., Montfort N. (edited by). 2003. *The New Media Reader*. Cambridge (MA): The MIT Press.

Wark, McKenzie. 2007. *Gamer Theory*. Cambridge (MA): Harvard University Press.

Weibel, Peter. 1992. *De nieuwe ruimte in het elektronische tijdperk*. In Bolle E.A. (edited by). *Boek voortinstabiele media*. Vol. 2, 65–75. Den Bosch (The Netherlands).

Weizenbaum, Joseph. 1993 (1976). *Computer Power and Human Reason: From Judgment to Calculation*. London (UK): Penguin Books Ltd.

Wells, Herbert George. 1983. *COMPLETE & UNABRIDGED* (collection of novels). London (UK): Heineman Octopus.

Westerlaken Michelle, Gualeni, Stefano. 2014a. "Grounded Zoomorphism: an evaluation methodology for ACI design". Paper presented at the ACE "14 Workshops in Funchal, Portugal, November 11–14 2014.

Westerlaken Michelle, Gualeni, Stefano. 2014b. "Felino: The Philosophical Practice of Making an Interspecies Video Game". Paper presented at the 2014 Philosophy of Computer Games conference, held at the Bilgi University of Istanbul, Turkey, November 13–16, 2014.

White, Carol J. 1996. "The Time of Being and the Metaphysics of Presence". In *Man and World*, 1996, Vol. 29, 147–166. Dordrecht (The Netherlands): Kluwer Academic Publishers.

Wilson, Catherine. 2000. *Vicariousness and Authenticity*. In Goldberg, Ken (edited by). *The Robot in the Garden: Telerobotics and Telepistemology in the Age of the Internet*. Cambridge (MA): The MIT Press. 64–88.

Wolf, M. J. P., Perron, B. (edited by). 2003. *The Video Game Theory Reader*. London (UK): Routledge. 67–86.

Yee, Nick. 2014. *The Proteus Paradox – How Online Games and Virtual Worlds Change Us—And How They Don't*. New Haven (CT): Yale University Press.

Cited Videogames

Games Inc. 2012. *Dyad* (video game). PS3. Playstation 3.

Artoon. 2002. *Blinx: The Time Sweeper* (videogame). Microsoft Xbox, Microsoft Games Studio.

Bioware. 2007. *Mass Effect* (videogame). Microsoft Xbox 360, Microsoft Game Studios.

Capcom. 2005. *Resident Evil 4* (videogame). Playstation 2, Capcom.

Double Jungle S.a.s. 2011–2013. *Gua-Le-Ni; or, The Horrendous Parade* (videogame). iPad and iPhone.

De Gier, Ismail, Nijman, Barbosa Dijkstra, Muller, Veer. 2011. *Glitchhiker* (videogame). Windows. Available online at http://www.glitchhiker.com. (accessed on May 28, 2014).

Ezra "XRA" Hanson-White. 2012. *Memory of a Broken Dimension* (videogame). Windows. Available online at http://www.datatragedy.com/wipmoabd/ (accessed on May 28, 2014).

Gualeni, Dini, Gomez Maureira, Shanchez Sarquiz. 2013. NECESSARY EVIL (videogame). Windows. Available online at http://evil.gua-le-ni.com. (accessed on May 28, 2014).
Intelligent Systems. 2006. *Super Paper Mario* (videogame). Nintendo Wii, Nintendo.
La Molleindustria. 2007. *Operation Pedopriest* (videogame). Windows.
La Molleindustria. 2008. *Oiligarchy* (videogame). Windows.
La Molleindustria. 2009. *Everyday the Same Dream* (videogame). Windows.
Japan Studio. 2008. *Echochrome* (videogame). Sony PSP, Sony Computer Entertainment Japan.
Jonathan "Cactus" Söderström. 2009. *This is Infinity* (videogame). Windows.
Jonathan "Cactus" Söderström. 2008. *Psychosomnium* (videogame). Windows.
Jonathan "Cactus" Söderström. 2007. *Mondo Medicals* (videogame). Windows.
Marc ten Bosch. (In development). *Miegakure* (videogame). Windows.
Maxis. 2000 – present. *The Sims* (videogame). Electronic Arts. Windows.
MIT Lab. 2012. *A Slower Speed of Light* (videogame). Windows. Available online at http://www.freeindiegam.es/2013/02/a-slower-speed-of-light-mit-game-lab. (accessed on May 28, 2014).
Nintendo R&D1. 1984. *Duck Hunt* (videogame). Nintendo Entertainment System, Nintendo.
Nintendo R&D4. 1985. *Super Mario Bros.* (videogame). Nintendo Entertainment System, Nintendo.
Nintendo EAD. 1996. *Super Mario 64* (videogame). Nintendo 64, Nintendo.
Nintendo EAD. 2000. *Legend of Zelda: Majora's Mask* (videogame). Nintendo 64, Nintendo.
Nintendo EAD. 2006. *Legend of Zelda: Twilight Princess* (videogame). Nintendo Wii, Nintendo.
Number None. 2008. *Braid* (videogame). Microsoft Xbox 360, Microsoft Game Studios.
Q Entertainment & HexaDrive.(2001). *REZ* (videogame). Sega Dreamcast. SEGA.
Quantic Dream. 2010. *Heavy Rain* (videogame). Playstation 3, Sony Computer Entertainment.
Raven Software. 2009. *X-Men Origins: Wolverine* (videogame). Microsoft Xbox 360, Activision.
Rocksteady Studios. 2009. *Batman: Arkham Asylum* (videogame). Microsoft Xbox 360, Eidos Interactive.
Sega AM2. 2000. *Shenmue* (videogame). Sega Dreamcast, Sega.
Tale of Tales. 2012. *Bientôt l"été* (videogame). Windows.
Technically Finished. 2009. *Haerfest* (videogame). Windows.
UbiSoft Montreal. 2003. *Prince of Persia: The Sands of Time* (videogame). Playstation 2, Ubisoft.
Valve Corporation. 2007. *Portal* (videogame). Microsoft Xbox 360, Valve Corporation.
Valve Corporation. 2011. *Portal 2* (videogame). Microsoft Xbox 360, Valve Corporation.
Jarrad "Farbs" Woods. 2008. *ROM CHECK FAIL!* (videogame). Windows. Available online at http://www.farbs.org/romcheckfail.php. (accessed on May 28, 2014).
Zoë Mode. 2007. *Crush* (videogame). Sony PSP. Sega.
Zynga. 2009. *Farmville* (videogame). Web browser (Flash).

Index

Printed and bound by CPI Group (UK) Ltd, Croydon, CR0 4YY